T0176301

BALANCE

BALANCE

HOW IT WORKS AND WHAT IT MEANS

PAUL THAGARD

Columbia University Press *New York*

Columbia University Press
Publishers Since 1893
New York Chichester, West Sussex
cup.columbia.edu

Library of Congress Cataloging-in-Publication Data
Names: Thagard, Paul, author.
Title: Balance : how it works and what it means / Paul Thagard.
Description: New York : Columbia University Press, [2022] |
Includes bibliographical references and index.
Identifiers: LCCN 2021046717 | ISBN 9780231205580 (hardback) |
ISBN 9780231556071 (ebook)
Subjects: LCSH: Equilibrium. | Motion.
Classification: LCC QC131 .T43 2022 | DDC 531–dc23/eng/20211130
LC record available at https://lccn.loc.gov/2021046717

Columbia University Press books are printed on permanent and
durable acid-free paper.
Printed in the United States of America

Cover design: Philip Pascuzzo

To the heroes of the pandemic

CONTENTS

ACKNOWLEDGMENTS

This book was written during the COVID-19 pandemic and is dedicated to the medical personnel, scientists, engineers, and leaders who have kept it from being an even larger disaster.

For helpful suggestions, I am grateful to Chris Eliasmith, Laurette Larocque, Adam Thagard, and Dan Thagard. For valuable comments on an earlier draft, I thank Kate Anderson, Steve Bank, Jim Bauer, Rob DeSalle, Randy Harris, John Holmes, Lauren Talalay, and Joanne Wood. Anonymous reviewers provided some annoying comments that prompted improvements.

I have recycled some material from my *Psychology Today* blog, *Hot Thought*, for which I hold copyright.

Thanks to Miranda Martin, Kathryn Jorge, and Ben Kolstad for editorial support, Susan Zorn for copyediting, and ARC for the index.

Supplemental material including live web links can be found at paulthagard.com.

BALANCE

1

BALANCED BODIES AND LIVES

I took balance for granted until 2016, when I had a nasty bout of vertigo. I was getting up from the weight bench in my basement when the room started to spin wildly. I almost fell off the bench and could only walk upstairs by holding on to the railing and wall. The next day I could get around with a cane and saw my family doctor, who had me lie down and turn my head to the side. Dr. Wang noticed that my eyes were moving in an odd pattern called nystagmus, and she diagnosed me as having benign paroxysmal positional vertigo. This kind of vertigo is caused by errant calcium crystals in the inner ear and counts as benign because the crystals can be moved back in place by a head-positioning exercise called the Epley maneuver. Now, if I start to feel even a bit dizzy, I do the maneuver and vertigo is prevented.

My family members and friends have had other forms of vertigo. My son Adam was laid out for a week by extreme vertigo from a cold that developed into an ear infection. My other son Dan felt wobbly as one of the symptoms of a scary case of COVID-19. Many years ago, a friend of mine suffered from Ménière's disease, an ear disorder that causes extreme vertigo and nausea. Another friend has severe vertigo resulting from ministrokes that caused damage in his midbrain and cerebellum. Dizziness is among the most common problems that patients report to their doctors, and falling puts older people at major risk of fractures. What goes wrong in our balance systems to produce vertigo, dizziness, and falling?

In 2019 I saw that my local recreation center was offering courses in tai chi, an ancient Chinese form of exercise said to improve balance. I had seen tai chi in movies and thought it looked incredibly slow and boring. But I learned from the course that performing intricate routines at a controlled pace is challenging both physically and mentally. I was heartened by evidence that tai chi dramatically reduces falls in older people and has numerous other well-documented health benefits. I wrote a blog post that explained the effectiveness of tai chi in terms of cognitive neuroscience rather than the traditional Chinese medicine concepts of *yin*, *yang*, and *qi*.[1]

Thinking about balance, vertigo, and tai chi made me notice that varieties of balance are important in many domains. Biologists talk about the balance of nature and chemists determine if a reaction is in equilibrium. Economists debate whether budgets should be balanced and politicians worry about the balance of powers. The traditional visual symbol for the law is the balance scale, which is also the image for my astrological sign of Libra. People fret over whether they are eating a balanced diet and managing an appropriate work-life balance. Unbalanced minds suffer from maladies such as depression and anxiety. One of my favorite movies is Alfred Hitchcock's *Vertigo*. Reflective equilibrium is an influential concept in philosophy.

During the COVID-19 pandemic in 2020–2022, debates raged about how much public life and the economy needed to be locked down to prevent the spread of the disease. These debates concerned how to balance lives against livelihoods and health against prosperity, as shown in the politician's dilemma in figure 1.1. Health officials recommended strong measures such as closing businesses, but many political leaders tried to keep the economy functioning. Other conflicts in this balancing act included reconciling individual freedom with social controls such as requiring people to wear masks and to practice social distancing.

I became increasingly curious about how the concepts of balance and vertigo are used in diverse fields and how these uses are related to biological ideas from physiology and medicine.

FIGURE 1.1 Balancing lives and livelihoods in the COVID-19 pandemic, showing Premier Doug Ford of Ontario, Canada.

Source: Graeme MacKay/Artizans.com. Reprinted by permission.

Is there any connection between balancing the body and balancing the world, between physiology and society? This book begins with biological explanations of how the brain manages to balance the body and of how breakdowns in this management can lead to failures such as vertigo. A crucial question neglected by balance experts is why conscious feelings are associated with successful balance and especially with breakdowns such as dizziness and vertigo.

Literal kinds of balance, as well as failings such as vertigo, give rise to rich metaphors that pervade discussions of nature, medicine, society, the arts, and philosophy. I will show that the brain's ability to interpret many kinds of sensory inputs that influence balance is mirrored by the coherence and usefulness of many balance metaphors. Our minds expand on the body's balance to understand the rest of the world.

THE PUZZLE OF BALANCE

In 2012, Nik Wallenda walked a tightrope across Niagara Falls (figure 1.2). He had to deal with high winds, wet air, and a narrow rope, whereas ordinary people only have to walk along sidewalks and paths. But even ordinary walking has its perils, such as sidewalk cracks to step into and pathway roots to trip over. Nevertheless, people usually manage to make their way around the world without falling down or getting dizzy.

To keep us balanced, the brains of Wallenda and ordinary people have to solve an extraordinarily complicated puzzle. The pieces of the puzzle come from many sensory sources. Wallenda's eyes enable him to see the tightrope ahead and where his foot is landing on it. Just as important, his arms and legs tell him what is happening with the upper and lower parts of his body

FIGURE 1.2 Nik Wallenda crossing Niagara Falls.

Source: Dave Page/Wikimedia Commons.

and with the pole that he carries. Less obviously, sense organs in his inner ears detect the orientation and movement of his head and alert him if falling starts. Wallenda's senses send conflicting messages to an area of his brain stem that must come up with a coherent interpretation that makes sense of them. Analogously, when people do jigsaw puzzles, they have to put the pieces together in ways that are coherent with the shapes, colors, and content of the pieces, such as the pieces shown in figure 1.3.

In its pursuit of coherence, the brain stem is aided by other brain areas that send their own signals that are usually helpful but sometimes confusing. The cerebellum provides additional information

FIGURE 1.3 Pieces of a jigsaw puzzle that can fit together.

Photo by Paul Thagard.

and expectations about the location and motion of the body. The ocular cortex sends messages concerning the orientation and movements of the eyes, which are constantly reacting to what the body does. The thalamus and hippocampus furnish additional information about sensory inputs and spatial orientation that enables the brain to maintain balance and detect challenges to it. These brain areas and sensory information together produce an overall coherent picture of what is happening in the balanced body, whether it is walking a tightrope or just strolling along a street.

HOW BALANCE FAILS

Jigsaw puzzles can fail for various reasons. Sometimes pieces are missing because of manufacturing mistakes, and pieces can be damaged in ways that affect their color and shape matches. I gave up doing a challenging puzzle called "Bizarre Bookshop 2" after I had done the easy parts of edges and large objects because my progress in finding connections between almost identical pieces had become painfully slow. I just could not make further sense of it.

Similarly, solving the puzzle of balance can fail because of problems with the pieces and connections. Many balance problems, such as my benign form of vertigo, result from breakdowns in the parts of the middle ear that include fluid-filled canals and hairlike cells that detect motion in the fluid. Such broken parts send bad signals to the brain, which misinterprets them as saying that the head or the room is spinning.

Eye problems can also disturb balance when the two eyes fail to focus together or disagree about the size of images. Feet that have lost sensation because diabetes has affected the nerves can send signals to the brain that are inaccurate about movement. Damage to the cervical spine resulting from trauma, arthritis, or artery blockage can send misleading signals to the inner ear and brain, leading to dizziness and vertigo.

Other balance problems result from interactions among sensory pieces that do not fit together. Motion sickness can make you feel dizzy and nauseous from a mismatch between what your eyes and your inner ear are telling your brain. Damage caused by strokes can block the ability of important brain areas such as the cerebellum to integrate information from the senses and different brain areas. These brain areas then fail to communicate well with others, including the motor areas needed to prevent falls.

Such breakdowns are similar to what happens in social groups such as families and workplaces. These groups work well when each individual is functioning properly in turning perceptions into actions and when good communication among the individuals produces successful coordination and cooperation. But a group can fail because one or more of the individuals malfunction physically or mentally or because of failures in communication between them. Family breakdowns such as divorce occur from serious problems in family members and their inability to communicate with one another. Similarly, balance depends on the proper functioning of multiple senses and brain areas as well as on successful communication among them.

In medicine generally, good health depends on the operation of systems such as the heart and lungs. Diseases arise from breakdowns in the parts and their interactions, which can be treated by fixing these problems. For example, a heart attack results from a blockage in an artery that the heart uses to pump blood to the lungs and other parts of the body. Treatment requires fixing the blockage through drugs, stents, or surgery.

Similarly, treatment of vertigo and other balance problems requires repairing the parts and interactions that are not working as they should. The Epley maneuver that restored functioning to my inner ear is a successful example, but dealing with other kinds of vertigo, dizziness, and falling is more difficult. Unfortunately, treatments are scarce for Ménière's disease, stroke-induced vertigo, and other problems.

HOW BALANCE FEELS

When we walk, run, dance, or perform other feats of balance, we usually do not think about what we are doing. But when balance fails in falls, dizziness, or vertigo, we become intensely aware that something is wrong. A spinning room affects visual perception but also changes conscious experiences such as feelings of nausea and emotional reactions such as surprise and fear.

Astonishingly, the textbooks and articles I have read about balance do not even mention consciousness because they are mostly written by biologists rather than by psychologists or philosophers. Until recently, neuroscientists were leery of discussing consciousness lest they be accused of philosophizing and looking for God among the neurons. Fortunately, important ideas about the neural basis for consciousness can now be applied to balance and its disorders. These ideas include information integration, neuronal broadcasting, sensory encoding, and competition among neural representations.

A theory of consciousness can combine these ideas to provide unified answers to pressing questions. Why is normal balance mostly unconscious, whereas disorders such as dizziness, vertigo, and falling are unpleasantly conscious? Why are feelings of imbalance accompanied by emotional experiences such as surprise and fear? Why are different experiences associated with vertigo, such as feeling the room spin versus feeling the head spin? Why does balance influence consciousness at all? Whether we can answer such questions provides a strong test of the adequacy of competing theories of consciousness.

WHAT BALANCE MEANS

In 1930, Albert Einstein wrote to his son that people are like bicycles—they can only keep their balance when they keep moving.[2] Maintaining balance on a bicycle is the physiological process that requires the integration of sensory information from the eyes, inner ears, arms, and legs through the interaction of numerous brain areas. In contrast, Einstein's talk of balance in people is metaphorical because he means having a life that is balanced in goals and activities, not just staying upright. Balance metaphors pervade human discourse from physics to economics to philosophy; examples include *thermal equilibrium*, *bank balance*, and *balance as a key to the meaning of life*. Imbalance metaphors are also important in many domains, from *falling in love* to *tipping points* in climate change.

So to understand the meaning of balance in human lives, we need to examine the metaphorical use of concepts such as equilibrium and vertigo. Metaphors are implicit comparisons between a perplexing target and a source that illuminates it. The claim that life needs a balance between work and relationships illuminates the target of human lives, by comparing life to the source concepts of balancing the body or balancing objects on a weight scale.

Metaphors can serve many purposes, including explanation, advice, persuasion, and entertainment. The value of a metaphor depends on how well it accomplishes its intended goal. The most successful metaphor in the history of science is Darwin's concept of *natural selection*. Nature does not actually select the fittest individuals, but the comparison between competition among species and the selective actions of animal breeders contributed to many of the insights in *On the Origin of Species*. The explanatory contributions of natural selection are still central to modern biology.

But not all metaphors are so successful, and we can evaluate them as strong, weak, bogus, or toxic. Strong metaphors are ones like *natural selection* that fully accomplish their goals, whereas weak metaphors make only a small contribution to what they are supposed to accomplish. The metaphor of dealing with cancer as a battle or struggle can be helpful in encouraging patients to get good treatments, but it has harmful social effects when it suggests that a patient has failed to try hard enough to get better.[3]

Bogus metaphors are ones that might sound useful but in fact fail completely in accomplishing their intended goals. Creationists use the metaphor of *intelligent design* to try to explain the origins and complexity of the universe, but this hypothesis has been surpassed in explanatory power by physics and biology. Astrologers use signs such as Taurus as metaphors for personality and fate, without any empirical basis. Creationist and astrological metaphors are bogus because they totally fail at their implied goals of explanation and prediction, even if some people find them comforting or amusing.

Even worse than bogus metaphors are toxic ones that directly cause harm. The most toxic metaphors in the history of humanity are racist and sexist ones that are too offensive to list. Most discussions of metaphors assume that they make valuable contributions to language and thought, but the evil impact of racist and sexist metaphors highlights how metaphors can be dangerous.

Bogus medical metaphors such as *body energy* found in complementary medicine also qualify as toxic when they lead to ineffective treatments for diseases that could be cured by evidence-based methods. The most toxic balance metaphor I have found claims that vaccines are unhealthy because they disturb the body's natural balance.

My analysis of balance metaphors across many domains unflinchingly identifies them as strong, weak, bogus, or toxic. I begin with the study of the natural world, which benefits from strong metaphors such as *chemical balance* and *thermodynamic equilibrium*. More surprisingly, I will argue that the popular metaphor of the *balance of nature* is bogus because it assumes a degree of stability that is rare in biological systems. More useful is the imbalance metaphor of *tipping point* that highlights the dynamics of rapidly changing ecologies.

Modern medicine employs strong balance metaphors such as *balanced diet* and *immune system balance*, but ancient medicine used bogus metaphors that claimed that diseases result from imbalances in humors (Greek), yin/yang (Chinese), or doshas (Indian). When used today as a replacement for scientific medicine, these metaphors qualify as toxic because they direct people away from more effective treatments.

Balance metaphors abound in the social sciences, but most of them have weak explanatory power. Popular psychology uses vague notions such as a *steady relationship*, while scientific psychology has tried to develop richer balance theories such as *cognitive dissonance*. Theories of mental illness based on *chemical imbalances* have been revealed as simplistic. Some economic metaphors such as *balanced budget* are moderately useful, but the concept of *economic equilibrium* has sometimes blocked theoretical and practical progress. Political metaphors such as *balance of power* are useful but cry out for deeper explanations. Balance metaphors such as the *scales of justice* are important in other social areas, including law, journalism, sports, and food.

Balance metaphors operate in the artistic fields of literature, film, painting, and music. I show the artistic contribution of balance metaphors by analyzing the novel *A Fine Balance* and the 1958 movie *Vertigo*. Balance metaphors also operate in paintings that employ symmetry to generate beauty and in music that employs harmony for emotional effect.

Finally, balance metaphors have made valuable contributions to philosophy in helping us understand the nature of knowledge, rational decision making, morality, and the meaning of life. I introduce a new concept, *metabalance*, that concerns the problem of finding a balance between balance and imbalance, both of which can be valuable in some circumstances. Balance is usually the desired state, but imbalances can contribute to many projects, from architecture to entertainment.

Most of the metaphors I discuss are embodied in that they build on representations derived from our senses, including the bodily sense of balance and physical orientation. But metaphors also can go beyond the senses, as in the theological ideas used in the ancient Egyptian *Book of the Dead*. Figure 1.4 depicts

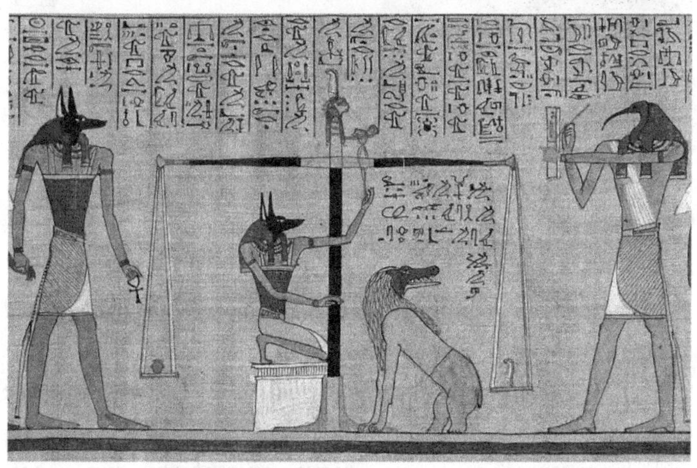

FIGURE 1.4 Weighing the soul, from the Egyptian *Book of the Dead*.

Source: Wikimedia Commons.

gods with animal heads weighing a dead man's soul against the "feather of truth" to determine if the man is worthy of an afterlife. Because ideas about gods, the soul, and truth go far beyond the body's senses, we need to understand how the mind's cognitive operations can abstract beyond the body.

A major difference between ancient and modern metaphors in science and medicine is that modern metaphors point to mechanisms that indicate how parts interact to produce changes. For example, Darwin's idea of *natural selection* has blossomed through the development of much more specific theories in population genetics. Similarly, the concept of *immune system balance* draws on a rich account of how antibodies protect but sometimes also threaten the functioning of organs. Metaphors can inspire and mingle with mechanisms.

Even without mechanisms, strong metaphors need to achieve coherence in that the correspondences between target and source concepts need to make sense. Balance in life corresponds systematically to balance on a bicycle because in both cases people detect and control competing forces to make progress. Bogus metaphors such as disease as *energy imbalance* fail because the correspondences do not work as intended.

MAKING SENSE OF BALANCE

So coherence is important for both literal, physical balance and for the many kinds of metaphorical balance. Keeping physical balance requires coherence among sensory signals from the eyes, inner ears, and limbs and among interacting internal signals from several brain areas. Similarly, productive use of balance metaphors requires coherence between the puzzling target and the illuminating source, including satisfaction of the goals of the comparison. I will use the term "sensemaking" for the brain's ways of coming up with coherent interpretations.

This book describes how the brain balances the body and why failures sometimes result in vertigo, falls, and nausea. It breaks new ground by explaining how balance and imbalance generate conscious experiences. Balance metaphors illuminate all areas of human life, from life-work balance to tipping points in climate change. What unites the physiological and metaphorical uses of balance is that they enable people to make sense of events by solving puzzles about perception and action.

But what is coherence? We shall see that both physical balance and metaphorical balance are achieved by satisfying constraints that determine what fits together and what does not. Both physiological and metaphorical balance require making sense, which usually works but sometimes fails badly, generating fall-inducing vertigo or toxic metaphors.

Understanding balance in both its literal and metaphorical senses requires a shift in how most people think about thinking. We tend to view inference as a series of verbal statements, as in Descartes's famous "I think, therefore I am." But sensemaking for balance requires nonverbal brain processes that include sensations, perceptual images, and emotions.[4] The brain has billions of neurons that operate in parallel—all at once—in contrast to the serial, step-by-step operations of conscious language. Sensemaking solves puzzles about balance by simultaneously using many kinds of representations and constraints to interpret what is going on in our bodies and our lives.

I have tried to make this book accessible for a general readership while contributing to neuroscience, psychology, and philosophy. Early chapters connect balance and vertigo with leading ideas in theoretical neuroscience, including the role of consciousness. Later chapters enrich the psychology of metaphor and analogy by analyzing the structure and impact of fifty-seven balance metaphors. The main philosophical contributions are normative evaluations of the strengths and weaknesses of the balance metaphors and arguments that some balance metaphors should be replaced by more exact descriptions of coherence construed as constraint satisfaction. Sensemaking illuminates both how balance works and how metaphor works. Moreover, just as bodily balance sometimes fails in disorders such as vertigo, balance metaphors sometimes fail in expressions that are bogus or toxic. Making sense of balance enhances our understanding of bodies, brains, and the human condition.

2

BALANCE AND THE BRAIN

When you walk down the street, you aren't aware of all the work that your brain is doing to keep you moving. If the pavement is uneven, you usually don't trip and fall down but make a quick adjustment to regain your balance. Moreover, you see the trees and houses as moving smoothly past you even though your head is moving up and down with each step. But if you stumble seriously and fall toward the ground, you stop thinking about all the other things on your mind and feel afraid as the pavement approaches.

A theory of balance should explain how you can do all this by answering these questions:

1. Why don't you fall over when you stand, walk, or climb a ladder?
2. Why doesn't your visual world bounce around when you move?
3. Why is balance sometimes ruined by disorders such as vertigo?
4. Why are you largely unaware of balance until you lose it when dizzy or falling?

This chapter answers the first two questions, leaving disorders for chapter 3 and consciousness for chapter 4.

The English word "balance" comes from French and Latin words for scales used in weighing, but balance has been around much longer than weight scales. Vertebrates such as fish, reptiles, birds, and mammals have similar balance mechanisms, and this sense of balance is more than 400 million years old. In contrast, weight scales originated in ancient Egypt and the Indus Valley around 5,000 years ago.[1] Cats do not have words for balance but have excellent mechanisms, as shown by their ability to right themselves in the air when falling. Animals such as chimpanzees have the balance needed to climb and move through trees, but no one has ever observed an ape holding two objects to determine which is heavier. So control of the body is much more fundamental to balance than weighing things on a scale.

Other concepts related to balance also have words connected to weighing. The word "equilibrium" comes from Latin roots meaning "equal" and "scales," and the more obscure "equipoise" means equal weight. One meaning of the word "stability" is physical and mental equilibrium. Biologists use the term "homeostasis" to mean stability among independent elements, such as the equilibria found in living organisms that maintain a balance in fluids and energy.

All balance concepts have opposites, such as imbalance, disequilibrium, and instability. These concepts of balance and imbalance have widespread metaphorical applications to nature and society.

MECHANISMS AND BREAKDOWNS

In biology, mechanisms are combinations of connected parts that interact to produce regular results.[2] For example, breathing works through parts such as the windpipe, lungs, and blood vessels: air passes through the windpipe to the lungs, where blood vessels acquire oxygen. The result is oxygenated blood that can fuel other organs. Similarly, to explain how balance works, we need to identify the inner ear, eyes, body, and areas of the brain to specify how the interactions among these parts produce the stability of bodies and vision.

Mechanisms explain how balance works and how it sometimes fails in problems such as vertigo. Figure 2.1 shows a familiar machine, a pop-up toaster, which consists of connected parts that include the handle, slots, and wires, whose interactions produce the desired result of toasted bread. Figure 2.2 is a diagram of how toasters work, with parts (e.g., slots and handles) shown by ovals and interactions (e.g., putting bread into the slots) shown by rectangles. This picture is simplified because it does not include other important factors, such as the hand that puts the bread into the toaster's slots or the electrons that go through the cord to heat the wire that toasts the bread. Nevertheless, the

FIGURE 2.1 A pop-up toaster.

Source: Photo by Piotr Siedlecki.

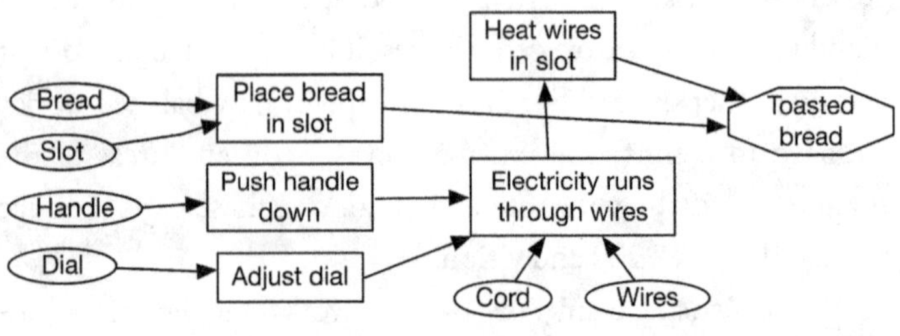

FIGURE 2.2 Mechanism for a pop-up toaster. Parts are indicated by ovals and interactions by rectangles. The octagon indicates the resulting change. Arrows indicate causality.

diagram highlights the main parts and the interactions that lead to toast.

The picture of the toasting mechanism shown in figure 2.2 points to important ways in which toasting can fail to work. Maybe the wires have no electricity because the cord is not plugged into the electrical outlet. Or maybe a loose connection in the wires prevents the electricity from heating them.

A mechanism can fail in four ways:

1. Problems with the parts, such as a broken handle or dial, or a dial set too high so that the toast gets burnt
2. Problems with the connections between parts, such as a slippery hand that does not get the handle down
3. Problems with the interactions between the parts, such as when a too-thin piece of bread falls under the slots and does not get toasted
4. Problems with interactions among the interactions, such as when the amount of electrical current through the wires is insufficient to heat the wires hot enough to toast bread

The balance disorders in chapter 3 illustrate all these ways in which a mechanism can break down.

THE INNER EAR

The internal sense of balance is less familiar than the external senses of vision, hearing, touch, taste, and smell. The external senses all have obvious organs that interact with the world: eyes, ears, skin, tongue, and nose. These organs produce nerve signals that go to the brain, where they are processed into perceptions. Internal senses that pick up signals from the body rather than the outside world include pain, heat, hunger, thirst, acceleration, time, stomach distention, proprioception/kinesthesia (body position and movement), and the need to urinate or defecate. External and internal senses use cells that detect changes in the world and generate signals, such as when the rods and cones in the retina are stimulated by light to excite nerves that connect to the back of the brain for processing into images.

For balance, the most important sense organs are parts of the inner ear, although interactions with vision and bodily senses also matter.[3] Ears have three parts as shown in figure 2.3: (1) beyond the visible outer piece is a canal leading to the eardrum, which vibrates when sound waves hit it; and beyond the eardrum is the (2) middle ear, a cavity with three small bones that transmit sounds from the air to the fluid in the (3) cochlea in the inner ear. Hearing is a mechanism that connects sound waves to the eardrum and cochlea, which has sensory cells that send signals to the brain for processing into auditory perceptions. Deafness can arise from various breakdowns in this mechanism, including damage to the eardrum or cochlea, blockage of ear canals by wax, damage to the nerves connecting the ear to the brain, and damage to the brain areas needed for processing signals from the ear.

The area of the inner ear except for the cochlea is called the vestibule, which is why the balance system is also called the vestibular system. To maintain balance, your brain needs to be informed about the motion and acceleration of your head.

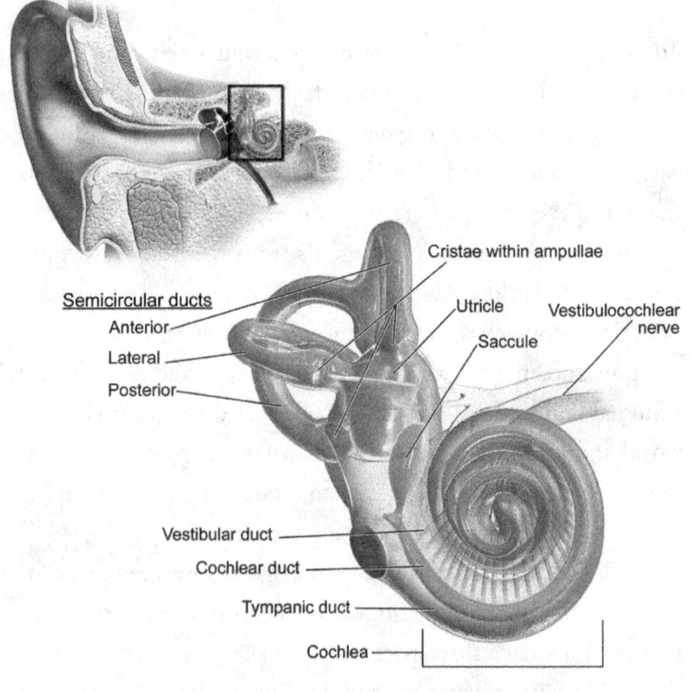

FIGURE 2.3 Structure of the inner ear. The semicircular ducts are more usually called canals. The vestibulocochlear nerve is also called the vestibular nerve.

Source: Blausen.com, "Medical Gallery of Blausen Medical 2014," *WikiJournal of Medicine* 1, no. 2 (2014).

Rotation can be in three directions: up and down, backwards and forwards, and left and right. Acceleration means speeding up or slowing down. Together these yield various possibilities, such as the head moving up with increased speed or the head moving left with decreased speed, all of which are tracked by sensors in the inner ear. In addition to affecting balance, the vestibular system controls eye movements and orients the body in space. By detecting head motion, the vestibular system also provides a sense that the body is moving.

Figure 2.3 shows the parts of the inner ear with the required motion sensors. The three semicircular canals (also called ducts) are filled with fluid that moves when the head moves. The three canals have different orientations, so the fluid in each one moves differently depending on which direction the head is rotating. Compare a carton of milk: the milk moves differently depending on whether you shake the carton up, left, or forward.

Unlike a milk carton, the semicircular canals contain cells that respond to the movement of fluid. The cells that do the sensing are called *hair cells*, although they have nothing to do with the kind of hair that you have on your head.[4] Hair cells consist of projections called stereocilia that are affected by fluid motion and bottom parts that generate chemicals. These chemicals stimulate neurons to fire, producing a neural signal in the vestibular nerve that sends information about head movement to the brain. The moving fluid in the semicircular duct pushes the hairs to generate signals that activate neurons.

This mechanism is similar to how hearing works in the cochlea, which also has hair cells. After sound waves vibrate the eardrum, the three small bones in the middle ear amplify the vibrations and transmit them to the fluid in the cochlea, where hair cells detect them and send neural signals to the brain. As a rough analogy, when you stick your hand out the window of a car and feel the wind with your fingers, you can sense both the speed and direction of the car's motion.

The function of the canal fluid and hair cells is to detect the motion of the head. Figure 2.4 displays the mechanism that translates head rotation into neural signals. The moving head affects fluid in the three semicircular canals, stimulating neurons to send signals to the brain. Moving the head in a particular direction,

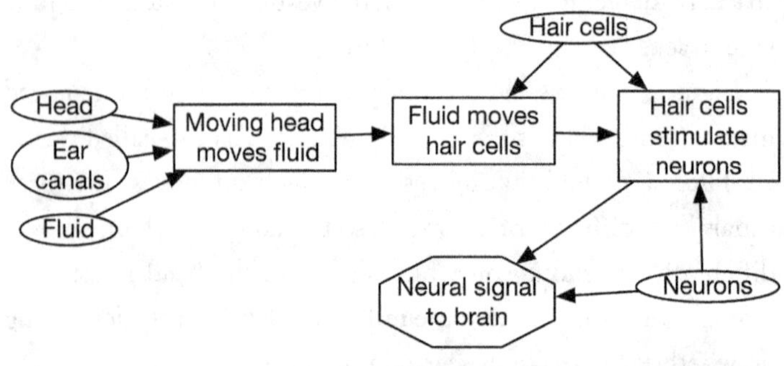

FIGURE 2.4 Mechanism for detection of head rotation. Parts are indicated by ovals, interactions by rectangles, and the result by an octagon. Arrows indicate causality.

such as rotating it to the left, has a different effect in the left and right ear canals than moving the head in other directions. Motion detection can go wrong if there is a problem with the fluid, perhaps as the result of an ear infection that makes the ear send signals to the brain that do not accurately reflect the head's rotation.

Acceleration is detected by two parts called otoliths (from Greek words for ear and stone): the utricle and saccule are shown in figure 2.3. Otoliths have crystals made of calcium carbonate, a substance also found in limestone and some antacids. Like semicircular canals, the otoliths contain fluid that is affected by the motion of the head, but their hair cells are different in that they have crystals on their ends. These crystals enable the projections to detect changes in velocity, similar to when you are in a car and the driver's speeding up pushes you back into the seat.

Because of their orientation, the utricle's crystals are sensitive to changes in horizontal movement, while the saccule's crystals are sensitive to vertical motion and gravity. Accelerated movement of the crystals stimulates the hair cells to release chemicals to initiate neural firings. Figure 2.5 shows the mechanism

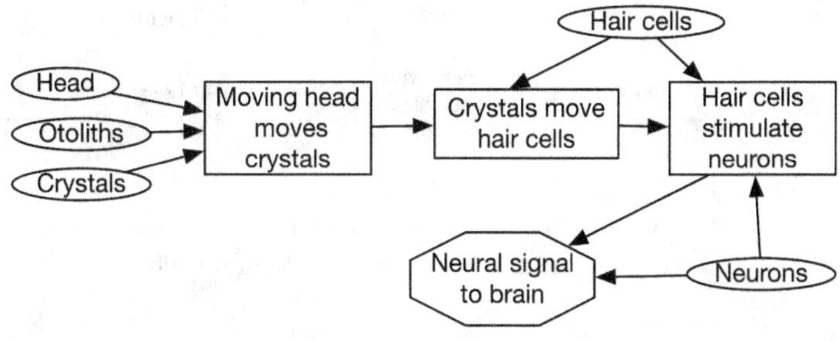

FIGURE 2.5 Mechanism for otolith detection of acceleration. Ovals are parts, rectangles are interactions, and the octagon is the result. Arrows indicate causality.

by which the inner ear detects acceleration through interactions among the crystals, hair cells, and neurons. This mechanism can break down when crystals slip out of place, as in my benign paroxysmal positional vertigo.

Smartphones have components that work like the parts of the inner ear. They have an accelerometer that detects linear motion like the otoliths and a gyroscope that detects rotational motion like the semicircular canals. In addition, a magnetometer detects changes in magnetic fields that are probably beyond human sensing, although some birds may be able to detect these changes. My Apple watch has a fall detection system that can tell if I have had a bad fall by combining information from its accelerometer and gyroscope, drawing on data collected from real falls in a movement disorder clinic and elsewhere.[5] Falling involves both a rapid change in acceleration, which is detected by the accelerometer, and a change in orientation of the watch, which is detected by the gyroscope.

Figures 2.4 and 2.5 show the motion-detection mechanisms for one ear, but having two ears with two sets of detectors allows

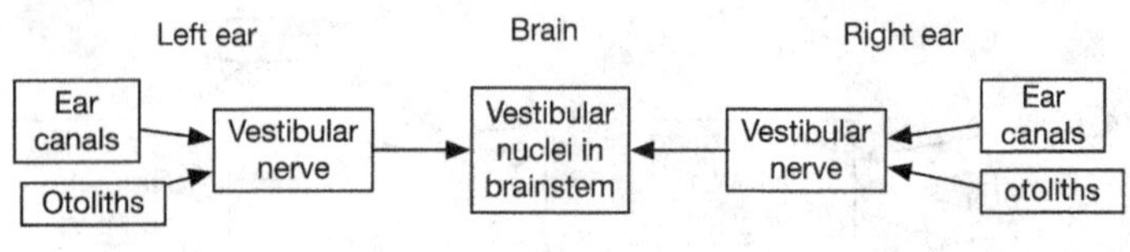

FIGURE 2.6 Convergence of information from the left and right ear on the brainstem. Arrows show flow of information.

for a richer estimate of the motion of the head, just as having sound detection in two ears helps you to capture directional information that one ear could not. Figure 2.6 shows the convergence on the brain of information from ten sources. Inner ears, which have powerful mechanisms for detecting motion in the head, send signals from six semicircular canals and four otoliths to the brain, specifically to the parts of the brainstem called the vestibular nuclei. The parts of the inner ear by themselves are not capable of performing the key balance functions of keeping the body upright and making vision stable. These accomplishments require interactions with many parts of the brain and body.

THE INTEGRATING BRAIN

The brainstem is located at the bottom of the brain just above the spine. A nucleus is a collection of neurons, and the vestibular nuclei are collections of neurons that integrate information from the inner ear with information from other brain areas. Signals go from the hair cell sensors in the canals and otoliths to the vestibular nuclei via the vestibular nerve, which is a bundle of axons that connect individual neurons in the inner ear with individual neurons in the nuclei.

When a neuron builds up enough of an electrical charge to fire, it sends an electrical signal down its axon, which has synaptic connections with other neurons. Synapses transmit chemicals that then excite or inhibit the firing of the recipient neuron. For vestibular neurons, the main neurotransmitter is glutamate, which is excitatory. The causal chain operates with neurons in the inner ear sending signals down the vestibular nerve to encourage neurons in the vestibular nuclei to fire. The job of the vestibular nuclei is to make sense of the flood of information it gets from the semicircular canals and otoliths.

The vestibular nuclei do not work with just the information from the inner ear because they have connections with other brain areas as well, as shown in figure 2.7.[6] All of the connections are reciprocal, with signals going from one brain area to another and back again. Connections also exist between the vestibular nuclei and organs in the inner ear such as the otoliths.

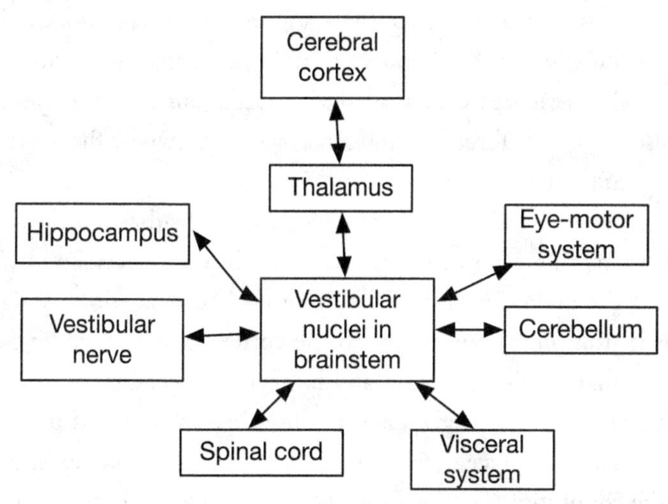

FIGURE 2.7 Interactions between brain areas. Arrows indicate neural signaling.

The bottom left of figure 2.7 illustrates the mutual connections with the spinal cord; these are important because the spinal cord's nerves carry proprioceptive information about the body and its movements, including the location and motion of the arms and legs. This information is important for balancing the body through reflexes and feedback loops. If the input from the inner ear says that your body is falling over, then you can move your legs and arms to regain balance. The interactions with the cerebellum are also important for control of the body.

Viscera are internal organs such as the stomach, and feelings of nausea often accompany issues with motion and balance. I cannot read a book in a moving car without feeling queasy because of interactions among the eyes, vestibular system, and stomach.

The connections between the vestibular nuclei and the eye-motor system are crucial for explaining visual stabilization when the body is moving. The inner ear detects motion and then informs the eye muscles via the vestibular nuclei to make compensating movements. These adjustments usually work well, but sometimes they produce eye movements that cause illusory visual experiences such as vertigo. Information from the eyes is also important for maintaining balance, as shown by the greater difficulty that some people have with moving in the dark.

The vestibular nuclei communicate with the thalamus, which is located on top of the brainstem and below the cerebral cortex. The main function of the thalamus is to relay sensory signals from numerous organs to the cortex, including vestibular information about motion and balance. The cerebral cortex in humans is much larger than it is in other animals and plays a major role in perception, memory, language, and consciousness. The connections between the vestibular nuclei and the cerebral cortex via the thalamus are crucial for explaining balance-related feelings such as dizziness and nausea.

Figure 2.7 also shows connections between the vestibular nuclei and the hippocampus, which keeps track of the body's location in space. Knowing where your body is helps to keep it balanced and vice versa. The hippocampus is also important for memory, which is also affected by the vestibular system.

For a small brain area found even in fish, the vestibular nuclei are amazingly powerful. They integrate information of many kinds, including the hair cells' detection of head motion, the eyes' detection of visual information, the body's detection of limb orientation, and even the stomach's detection of discomfort. They put all this information together to perform two key functions—maintaining the body's balance and stabilizing vision. In cooperation with other brain areas, the brainstem usually constructs a coherent interpretation of all the available sensory signals. When sensemaking fails, information about balance problems is transmitted to the cerebral cortex, where it can enter consciousness.

BALANCING MOVEMENT

The popular saying that cats always land on their feet turns out to be approximately true. Cats have a righting reflex that enables them to turn in midair and flatten themselves so that they can land safely even when falling from heights such as a tree. Humans have a less effective form of righting reflex that enables them to adjust their heads and necks when they are falling.

For both cats and humans, the balance system in the inner ear serves to correct posture because it detects dangerous motions such as a fall to the ground. Sometimes controlling posture and remaining upright is a simple reflex involving a signal from the inner ear to the vestibular nuclei to muscles that produce movements that correct the imbalance. But other times correcting balance is a much more complicated action performed by multiple interacting brain areas.

Reflexes

The mechanisms used in reflex corrections of balance have been identified through experiments on cats. In cats as in people, rotation of the head to the left causes motion in the fluid in the canals of the inner ear. Conversely, electrical stimulation of particular nerves in the cats' canals evokes head movements in the opposite direction so that, for example, apparent motion to the left causes the cat to turn back to the right.

The neural mechanism for such reflexes is straightforward, as shown in figure 2.8. Motion is detected in the inner ear by neurons that fire in response to signals from the canals. These neurons lead to firing in the vestibular nuclei of neurons that are directly connected with motor neurons in the neck. So inner ear canal detection of motion leads directly to compensating movements of the neck.

Similar reflexes operate when acceleration is applied to the head, such as when it is pushed forward by a shove. Such accelerations are detected by the hair cells in the otolith organs, which send signals to the vestibular nuclei, whose neurons have direct connections to neurons that control movements in the neck

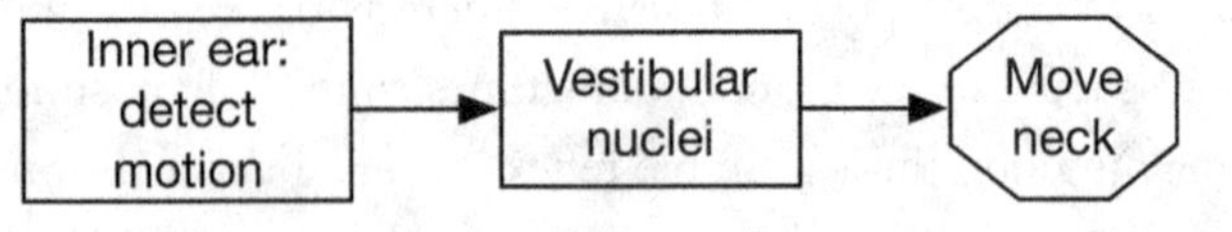

FIGURE 2.8 Reflex mechanism for moving neck to adjust for motion.

muscles. This process allows the balance system to compensate for head tilts by moving the neck.

The limbs also operate with motion-driven reflexes. If you feel that you are falling, you may adjust your ankles and hips or throw your arms up as a counterbalance. As with the neck muscles, direct links connect the neurons detecting motion in the inner ear canals and otoliths with motor neurons that control limb motions. So the brain is well equipped to make muscular adjustments to maintain motion without any conscious awareness of the process. Reflex mechanisms can fail because of problems with the inner ear, vestibular nuclei, or motor areas, or because of problems with interconnections between them. Falls may result.

Inferences

If someone throws a rock hard at your head, you automatically duck. But if the projectile coming toward your head is a soft ball, you have a variety of choices. You could knock it away with something that you are holding, or you could try to catch it with one hand or two. Human brainpower means that we do not always have to act on reflex. The brain can employ more complicated strategies that take into account multiple senses such as vision and body location as well as the information in the inner ear. Such strategies can produce activity in various parts of the body and can be adapted through experience to become reliable. By making inferences from different sensory inputs and motor strategies, we achieve flexible balance control.

Suppose you are standing on a rock that starts to move forwards or backwards. Your inner ears detect motion that your brain interprets as swaying or falling, but your body can regain equilibrium by combining different behaviors, such as stepping forward while raising your arms. Which of these behaviors is best depends on various factors, including how much the rock is moving, how slippery it is, what you are seeing, and what you expect to happen. An immediate reflex may keep you from falling off the rock, but a more complicated inference that combines all the information available to you may provide a better mixed strategy for maintaining your balance.

Reflexes can be captured by simple rules like *if falling to the left then move head to the right.* But balance problems require integrating information from the otoliths and the semicircular canals with signals from the eyes, body, and muscles. This integration generates an interpretation of the situation that could then lead to selection among several possible actions. With the slippery rock, your choices might be moving your hips to maintain balance on the rock or jumping to another rock that looks more stable. Sometimes the selection may even be conscious, as when you say to yourself: jump.

To perform these inferences about both interpretation and action, brain areas such as the vestibular nuclei construct an internal model of the situation.[7] Such models are not static objects such as a model airplane but rather dynamic representations of how configurations of sensory information lead to changing perceptions or novel actions.

Competing theories explain how groups of neurons model situations and actions. One approach supposes that the brain uses probabilities to assess different interpretations such as that you are falling off the rock and different actions such as stepping off the rock. I find these accounts implausible because mathematical probability is a powerful idea only a few hundred years old, whereas vertebrates have been balancing for hundreds of millions of years. More plausibly, neural mechanisms accomplish inferences that only look like probabilistic calculation from a distance.

Instead, we can think of internal models as being built through two simpler mechanisms: neural integration and parallel constraint satisfaction. Figure 2.9 shows the simplest kind of neural integration in which two neurons excite the firing of the third neuron so that the third neuron integrates information from the other two. Each neuron carries information by firing in a particular pattern, so that the firing pattern of the third neuron depends on the firing patterns of its inputs. As a simple example, suppose that the firing of neuron 1 represents *slippery* and the firing of neuron 2 represents *rock*. Then the integrated information captured by neuron 3 is *slippery rock*. More realistically, each of these concepts is represented by groups of thousands or millions of neurons. Then integration takes place when large groups of neurons interact to affect the firings of neurons in another group.

Figure 2.7 showed neural integration in the balance system, with neural groups in the vestibular nuclei putting together information from multiple sources such as the inner ear, eyes, and body. The mechanism of integration generalizes what the three neurons do in figure 2.9: neurons influence the firing of

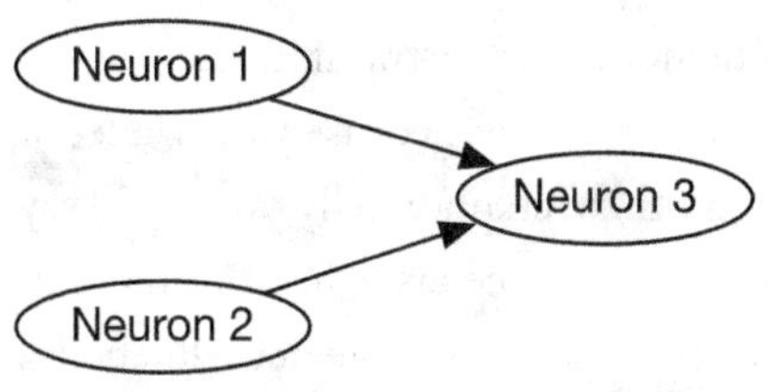

FIGURE 2.9 Neural integration with three neurons represented by ovals. The arrows indicate that firing of one neuron excites (increases the firing of) the recipient neuron. Alternative interpretation: ovals represent groups of thousands of neurons that are integrated by another neural group.

another neuron when they provide electrochemical signals that contribute to whether or not that third neuron is going to fire. You can imagine the vestibular nuclei as a government department that gets information from diverse sources that it has to put together to make decisions.

Neural integration provides a simple internal model that can work well if all information coming in is consistent. But perception often has to deal with inconsistent interpretations, such as when you cannot tell whether an object that you see in the sky is a bird or an airplane. Similarly, balance problems can be difficult if your inner ears and eyes suggest different interpretations, as happens with motion sickness when ear canals say that you are moving and eyes focused on a book say that you are stationary.

Neural networks deal with such problems by the method of constraint satisfaction, which is most easily explained through examples of decision-making. Suppose you are in a restaurant trying to decide what meal to order but you face a difficult choice between different dishes that are appealing for multiple reasons such as taste, cost, and health; these are the positive constraints on your decision. A sirloin steak might appeal to you as the tastiest choice, but it is also one of the most expensive items on the menu and contains too much cholesterol to be healthy. Alternatively, a veggie burger might be cheaper and healthier but lacking in taste. The negative constraint is that you do not want both a steak and a veggie burger because your wallet or your appetite cannot handle both. Metaphorically, you could say that the decision requires you to balance taste, health, and cost.

Figure 2.10 shows how a simple neural network can deal with these conflicts. The positive constraints reflect how much the goals of taste, health, and cost are satisfied by the eating choices, as indicated by the solid lines. The incompatibility of the two choices is indicated by the dotted line, reflecting that you cannot

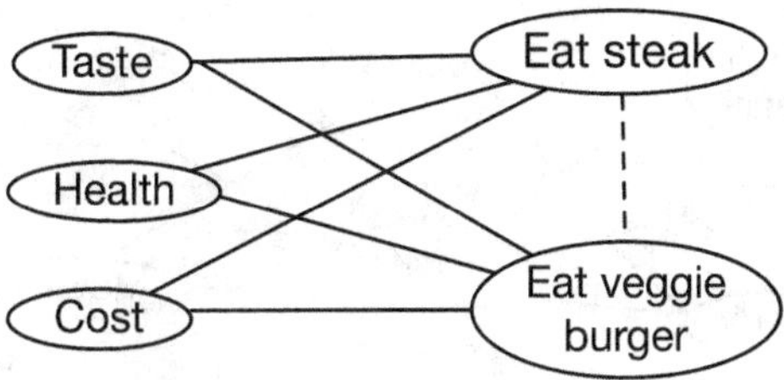

FIGURE 2.10 Decision-making about eating as a constraint satisfaction problem. Positive constraints are indicated by solid lines and the negative constraint by a dotted line.

have both the steak and the veggie burger because it is too much food and too expensive. Neural networks can implement these constraints by using neurons (or groups of neurons, or firing patterns in groups of neurons) to represent goals and choices, using excitatory links for positive constraints and inhibitory links for negative constraints. Simple algorithms adjust the firing rates of the neurons based on the links and the firing rates of connected neurons until the network makes a coherent decision.

Neural networks solve such problems because they have not just the excitatory connections shown in the neural integration in figure 2.9 but also inhibitory connections. When one neuron excites another, it tends to make it fire, whereas when one neuron inhibits another, it tends to make it not fire. Neural groups can similarly excite each other or inhibit each other.

A balancing situation such as standing on a slippery rock requires the solution of two constraint satisfaction problems, one about the correct interpretation of the situation and the other about deciding what to do. Figure 2.11 shows a neural network with components representing inputs from several sources—the canals and otoliths in the inner ear, the eyes, and the body senses. These support different interpretations of the current situation

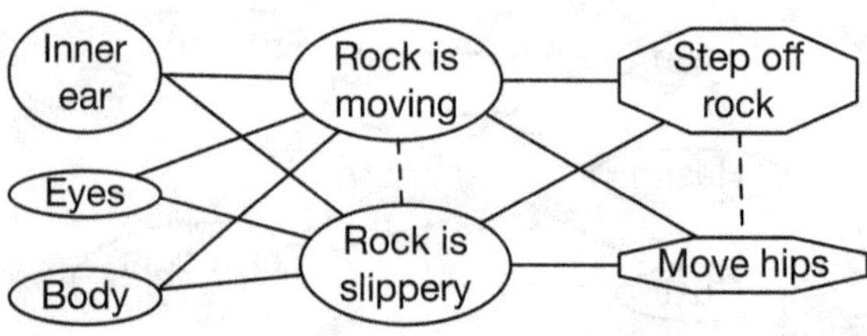

FIGURE 2.11 Rock balance problem requiring constraint satisfaction to interpret the situation and to figure out how to react to it. Positive constraints are indicated by solid lines and negative constraints by dotted lines.

that are incompatible, such as whether the rock is moving or just slippery. These interpretations compete, as shown by the dotted line in the center. Different interpretations support different compensating actions, such as moving hips or stepping off, which also compete (dotted line on the right). So neural networks can make important inferences and decisions, without recourse to probabilities and utilities, by practicing sensemaking through constraint satisfaction.[8]

BALANCING VISION

The balance system in your brain and body keeps you from falling over when you stand or walk, but it also solves another important problem. Vision begins by light hitting your retina, which sends nerve signals to the back of your head. But when you move, the pattern of retinal signals should change dramatically, which would turn your visual perception into an overwhelming jumble of images. The balance system prevents this jumble by a reflex that connects inner ear sensors with muscles that control the movements of the eye. The vestibular nuclei provide this connection just as they do for the reflex that enables the inner ear sensors to affect body movements in posture control.

The reaction that keeps visual images stable on the retina is called the vestibulo-ocular reflex, which moves the eyes to the right as the head turns to the left. During long head rotations, the eye reaches an extreme position and has to flip back to a new start. The resulting pattern of slow movements followed by rapid resetting is called nystagmus, which is heavily implicated in vertigo, discussed in the next chapter.

The mechanism for the vestibulo-ocular reflex is shown in figure 2.12. In the inner ear, the semicircular canals and otoliths detect head rotations. The vestibular nerve sends information about this to the vestibular nuclei in the brainstem. The neurons in these nuclei that receive this information connect to motor neurons that control eye movements. So movements of the head lead directly to motions of the eyes in the opposite direction to keep visual images stable.

The mechanism in figure 2.12 can be filled out by invoking more basic mechanisms. When the head rotates to the right, the fluid in each semicircular canal moves to the left, which has different effects on the left and right canals. The hair cells in the right horizontal canal get more stimulation and therefore send a bigger signal to the ventricular nuclei, whereas the hair cells in the left canal get less stimulation and send a smaller signal to the ventricular nuclei, which interprets the discrepancy as head

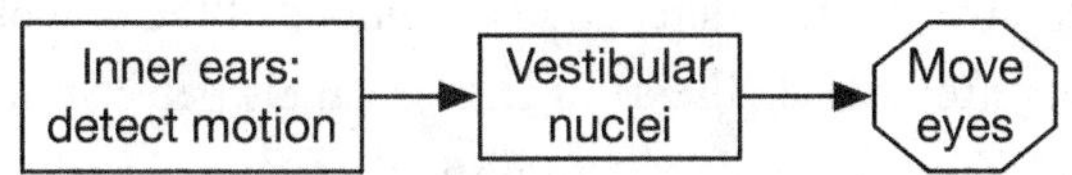

FIGURE 2.12 Vestibulo-ocular reflex, which stabilizes vision when the head is moving.

movement to the right. To compensate, the ventricular nuclei send signals to the eye muscles to move the eyes to the left. The three semicircular canals in each ear excite different eye muscles. Other reflexes react to signals from the otoliths responding to changes in head orientation relative to gravity.

CONTROLLING BALANCE

Balance is crucial in many activities besides just standing and walking. Enjoyable activities like sports and dance require a high degree of balance that depends on coordination of bodily movements carried out through the neural interactions shown in figure 2.7.[9] Skilled motions such as a basketball shot or a ballet pirouette expand the constraints to be satisfied beyond those required for simple balance sensemaking. The brain integrates proprioceptive information about bodily movement with vestibular information about balance to perform challenging moves.

The semicircular canals and otoliths in the inner ear of humans and all other vertebrates also contribute to high-level cognition, including navigation, spatial memory, and even self-consciousness.[10] Vestibular inputs influence object recognition, numerical cognition, and even social cognition, which aims at the perception and understanding of self and others.

More basically, this chapter has sketched mechanisms that enable the balance system to carry out the two important functions of maintaining posture and stabilizing vision. For both purposes, the motion-detection capabilities of the organs in the inner ear are crucial. The semicircular canals detect rotation in three dimensions, while the otoliths detect acceleration. These organs have hair cells that convert motion into nerve signals that are sent to the vestibular nuclei in the brainstem, where the different signals are interpreted as signs of motion. The signals to the vestibular nuclei are enough to generate reflexes in the form of movements by the limbs and eyes.

For controlling posture, many other sources of information are useful. The vestibular nuclei are interconnected with other brain areas involved with the spinal cord, the stomach, the eyes, and hippocampus. Controlling posture is not just a simple neuron-to-neuron reflex, but requires inferences to interpret a situation and to select reactions to instability, such as grabbing a handrail rather than shifting hips. These inferences are based on constraint satisfaction rather than on reflexes that use simple rules.

These mechanisms explain how balancing works to control posture and stabilize vision, indicating how sensemaking in the brain solves the puzzle of balance. Breakdowns in these mechanisms lead to the failures seen in harmful episodes such as falls and vertigo. Fixing the breakdowns provides treatments that reduce harm.

3

VERTIGO, NAUSEA, AND FALLS

When I was a child, my brothers and I liked to spin around until we got dizzy, staggered, and fell down. We didn't know that we were messing with our inner ear canals. Machines are wonderful when they work to accomplish their purposes, but all machines eventually break down. Your car drives you around until failures occur such as a flat tire or a dead battery. Your toaster makes bread taste better until a broken handle or a loose wire stops it from working.

Similarly, the balance system usually does a fabulous job of keeping you upright, enabling you to walk over rough terrain and keeping your eyes focused on important parts of your environment. But balance can fail as a result of breakdowns in crucial parts such as the sensors in the inner ear or in interactions such as processing in the brainstem or cerebellum. These breakdowns lead to disturbing experiences, including dizziness, vertigo, and nausea. Most seriously, loss of balance can lead to falls that cause broken bones or concussions.

To explain these disorders, we need to identify the breakdowns in parts and interactions that interfere with the normal functioning of biological mechanisms.[1] This pattern of explanation is standard in medicine, as table 3.1 illustrates. The thousands of human diseases include many whose causes are unknown, but known

TABLE 3.1 Bodily mechanisms whose breakdowns lead to diseases

System	Mechanisms	Breakdowns	Diseases
Cardiovascular	Heart pumps blood through arteries and veins.	Weakened heart muscle. Blocked arteries.	Congestive heart failure. Heart attack.
Respiratory	Lungs receive air through windpipe and deliver oxygen to blood.	Blocked windpipe. Tumor in lung. Infected lung.	Suffocation. Lung cancer. Pneumonia.
Digestive	Stomach and intestines digest food.	Inflamed stomach. Hole in lining. Inflamed bowel.	Gastritis. Ulcer. Crohn's disease.

causes are mechanism breakdowns. Table 3.1 shows some diseases associated with three bodily systems. For each system, many other mechanisms, breakdowns, and diseases could be listed, and a similar analysis could be done for the muscular, endocrine, immune, nervous, reproductive, urinary, skeletal, and other systems. Systems also interact, as when oxygen from lungs gets pumped through arteries. Accordingly, we should be able to explain balance disorders by indicating the parts and interactions that go wrong to produce failures of sensemaking: the process that is supposed to produce coherent interpretations of sensory experience.

WHY DOES THE WORLD SPIN?

Balance problems are one of the leading causes of people's visits to doctors, accounting for around 5 percent of visits.[2] I begin with vertigo, an intense form of dizziness that involves feelings of spinning or whirling.[3] Some people feel their heads spinning, but in my case my head felt stable but the room was spinning right to left. Either way, vertigo is the experience of illusory motion: something seems to be moving even though it is not. What goes wrong in brains that generate such illusions?

Chapter 2 described how the brain takes sensory signals from inner ears and combines it with information from eyes and body to make inferences and adjustments that keep you balanced. Perception is sensation plus interpretation, with sense organs providing signals and brain processes figuring out what they mean.

Balance perceptions can go wrong in two main ways. First, if the inner ears are not working properly, then they send bad signals to the brain, which then makes mistakes because it gets wrong information. Many kinds of vertigo, including the benign positional kind that I experienced, are caused in this way. Second, even good signals can have bad results if the brain areas that interpret them are defective because of problems such as trauma, tumors, and strokes. Balance is then like governments that make bad decisions because they get faulty information or because they interpret valid information stupidly, or both. Both balance and government can fail because of bad parts or poor interactions.

The key question I want to answer is: How do breakdowns in sensing and interpretation lead to the experience of spinning? I leave to chapter 4 the explanation of why vertigo, dizziness, and nausea are *conscious* experiences. My immediate concern is how the brain makes an inference that something is spinning, either the head or the room. The room spinning is the more surprising illusion because heads do sometimes spin, as we see with ballet dancers or figure skaters, whereas physical locations rarely spin except in unusual situations such as merry-go-rounds. Both kinds of illusions need to be explained by identifying the breakdown in sensing and interpreting mechanisms that can lead the brain to make wrong inferences about what is moving.

Many kinds of vertigo have been identified by medical research. Vertigo is not a disease but a symptom that can be caused by different disorders. Balance specialist Jack Wazen defines vertigo as "the illusion that either you or the environment around you is spinning, rotating, rolling, rocking, or whirling."[4] In contrast, dizziness is the feeling of light-headedness, faintness, or giddiness and only amounts to vertigo when whirling or spinning occur. Eye and neck problems can also cause dizziness. A third balance problem discussed by Wazen is disequilibrium, the inability to maintain control of your posture and to remain on your feet.

VERTIGO FROM INNER EAR DISORDERS

Wazen estimates that at least 40 percent of complaints of balance disorders result from problems with the inner ear, and I will explain some of the most common disorders that can cause vertigo. Chapter 2 describes how movement in the fluid in the semicircular canals can lead the brain to infer that the head is moving. In vertigo disorders, something goes wrong so that the brain infers that the head or room is persistently moving.

Figure 3.1 shows a simplified version of the mechanism by which the inner ear canals imply that the head is rotating in a particular direction. When you rotate your head to the left, the fluid in the ear canals moves to the left, which makes the hair cells in the left canal bend left and send a stronger signal, while

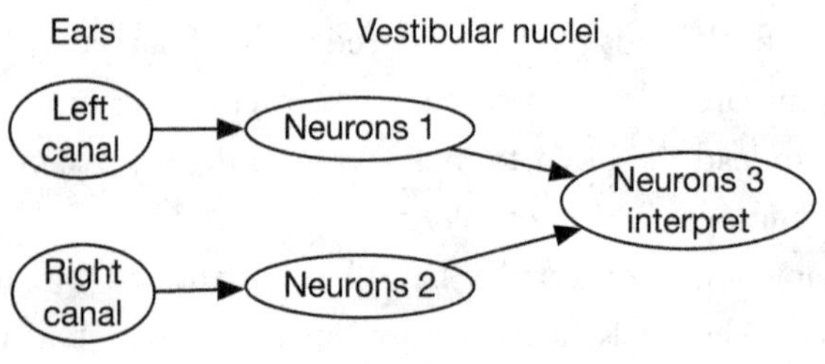

FIGURE 3.1 Mechanism for detecting that the head is moving in a particular direction. Ovals indicate groups of neurons, and arrows indicate excitation.

the hair cells in the right canal send a weaker signal. These hair cells send signals to groups of neurons in the vestibular nuclei labeled *neurons 1* and *neurons 2*. These groups of neurons connect with other groups labeled *neurons 3* that combine and interpret the signals coming from the two ear canals.

Ignoring other influences such as the otoliths, the rules for interpreting the signals amount to this:

> If the signal from the left canal is strong, then the head is moving left.
>
> If the signal from the right canal is strong, then the head is moving right.
>
> Otherwise, the head is not moving.

Here the signal is the rate at which the neurons fire or the pattern in which they fire. The hair cells generate a slow neural signal that increases when they are moved by fluid to bend in one direction and decreases when they are moved by fluid to bend in the other direction. The vestibular nuclei in the brainstem not only receive information from the inner ear but also modulate it because they have connections back to the hair cells.

Benign Paroxysmal Positional Vertigo (BPPV)

When I told my doctor that I had vertigo, she naturally suspected the most common kind, which is called "benign" because it is not life-threatening, "paroxysmal" because it occurs suddenly, and "positional" because it happens when the head's position rapidly changes.[5] I told her that my vertigo came on suddenly when I got up from my weight bench. Dr. Wang conducted the standard test by getting me to lie down on the examination table, rotating my head, and noticing that one of my eyes had nystagmus, a rapid, jerky movement back and forth generated by the vestibulo-ocular reflex.

Like most victims of BPPV, I had no head injury or acute infection, which sometimes initiates it. The nystagmus suggests that calcium carbonate crystals from my otoliths had somehow moved into my semicircular canals. Typically, the crystals from the utricle move into the posterior canal. The crystals moving in the ear canal provide extra stimulation to the hair cells sending the signal to the brainstem that the fluid is moving in the direction of that ear. If the crystals are out of place in your left ear, this leads the brain to think that the head is moving to the left. Unlike in a normal one-time rotation, the crystals continue to stimulate hair cells, implying that the head is constantly rotating. Alternatively, through connections to the visual system, the brain can infer that the room rather than the head is constantly moving.

Crystals out of place in the ear canal send a bad signal that the brain interprets as constant motion. The balance system fails because bad parts (ear canals disturbed by crystals) produce bad interactions with the brainstem, which interprets signals from them. The misinformation from the inner ear provides an illegitimate constraint that the brain can only satisfy by the misinterpretation that the head is moving.

Confirmation that I had BPPV came when the use of the Epley maneuver over a few days eliminated my vertigo. The Epley maneuver moves the head one way, then the other, and then down, pushing the crystals out of the canal so that hair cells no longer generate the false signal that the head is moving in that direction. As with toasters and other machines, fixing the bad part restores the mechanism. Medical treatments operate the same way, as when a stent fixes a blocked artery so that the heart can resume pumping blood.

Ménière's Disease

I have only had a few episodes of BPPV, but some people suffer from a different kind of vertigo for years. Ménière's disease is also a problem in the ear canal but has nothing to do with the displacement of calcium crystals. Rather, it results from excessive fluid buildup in the inner ear that affects both the semicircular canals and the cochlea, causing hearing problems such as deafness and tinnitus (ringing in the years). Figure 2.3 shows the relevant inner ear structure. Nausea, vomiting, and dropping suddenly to the ground can also be symptoms of the disease. Drop attacks are thought to be caused by a disturbance in the utricle otolith that disrupts the sense of gravity.

The accumulation of fluid in the inner year produces pressure in the chamber that holds the semicircular canals and the cochlea. Exactly how pressure induces vertigo is not known, but one possibility is that the pressure in the ear canals changes stimulation of the hair cells and sends a defective signal to the vestibular nuclei, much like the mechanism in figure 3.1. As with BPPV, this disease usually affects one ear, producing an ongoing signal to the brainstem, which does the interpretation: if the hair cells on one side are sending more nerve signals than the hair cells on the other side, then the head is moving in the direction of the side with more firings. Then excess fluid leads to excess nerve firing, which leads to the erroneous inference that the head is constantly moving.

Unlike BPPV, Ménière's disease is hard to treat because the cause of excess fluid is typically unknown. Ways of trying to reduce the amount of fluid include consuming a low-salt diet, taking a diuretic, avoiding caffeine and alcohol, stopping smoking, having surgery, and undergoing a program to reduce stress. Various drugs such as the antihistamine meclizine can help to prevent nausea and vomiting, and new drugs that can be delivered to the inner ear to reduce swelling are becoming available. Ménière's disease illustrates the difficulty of treating diseases whose causal mechanisms are unclear.

Labyrinthitis

The vertigo of my son Adam followed a cold that had turned into an ear infection, probably because the virus that had infected his throat and nose traveled through the Eustachian tube into his middle ear and then into his inner ear. An infection or inflammation of the inner ear is called labyrinthitis because the cavity that contains the semicircular canals, otoliths, and cochlea is called the labyrinth.

Most likely the mechanism by which labyrinthitis results in vertigo is similar to what happens in Ménière's disease. Inflammation in a canal on one side of the head makes that canal send an ongoing deviant signal to the vestibular nuclei, leading to the erroneous inference that the head is persistently moving. Trauma to the ear by a blow to the side of the head could have the same result.

Vestibular Neuritis

Viral infection can also cause malfunction in the vestibular nerve, which carries signals from the canals and otoliths to the brainstem. In vestibular neuritis, nothing is wrong with the signal coming from the semicircular canals, but it gets distorted because the infection of the nerve interferes with its ability to transfer the signal.

Alcohol

Before BPPV, my only experience with vertigo was from a youthful episode of excessive alcohol consumption. I would have thought that the episode resulted from the substantial effects of alcohol on the brain through neurotransmitter pathways that include dopamine, GABA, and opioids. Alcohol's enhancement of GABA, the major inhibitory neurotransmitter, may explain some motor effects such as stumbling and slurring words. But alcohol also has a direct effect on the inner ear.

Each semicircular canal has an ampulla that contains the hair cells that detect the fluid motion relevant to head movement. The ampulla has a structure called the cupula that is important because the hair cells are embedded within it. Alcohol diffuses from blood vessels into the cupula, making it buoyant because alcohol is less dense than the fluid in the canals. This buoyancy makes the cupula more sensitive to gravity, leading the hair cells to respond more to movements of the head. Then the canal misinforms the brainstem that the head is moving, leading to dizziness and vertigo.

BPPV, Ménière's disease, labyrinthitis, vestibular neuritis, and alcohol are just some of the ways that breakdowns in inner ear mechanisms can cause vertigo. Breakdowns in brain mechanisms cause other kinds of vertigo.

VERTIGO FROM BRAIN DISORDERS

Perception is not a simple pathway from sensation to interpretation because knowledge stored in different brain areas can contribute to the interpretation. For example, vision isn't just a matter of light hitting your retina and generating a perceived image because your previous experience stored as images and concepts can also influence what you perceive. Sometimes the mingling of sensation and interpretation can generate visual illusions such as the goblet/faces image shown in figure 3.2. Whether you see the figure as a goblet or as two faces depends on your past experience with goblets and faces and what you are expecting to see.

FIGURE 3.2 Ambiguous image that can be viewed as either a black goblet or two white faces.

Source: Wikimedia Commons.

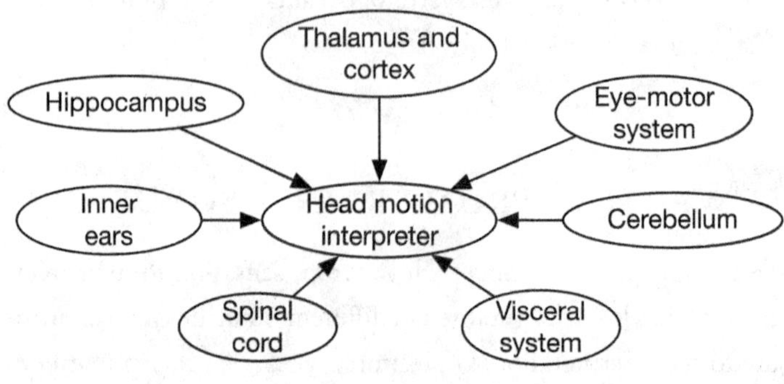

FIGURE 3.3 Influences on the neurons in the vestibular nuclei that infer head motion. Ovals indicate neural groups, while arrows indicate effects of neural firings.

Similarly, how you interpret your head's motion can depend on previous experiences saved in any of the brain areas that are part of the balance system. Figure 3.3 shows brain areas that affect how the vestibular nuclei interpret the signals coming from the inner ear. So, if those brain areas such as the cerebellum are defective, they can send signals that produce an erroneous interpretation that the head is in motion, generating the experience of dizziness or vertigo. Kinds of brain defects that can generate abnormal neural signals include strokes, tumors, multiple sclerosis, and aging.[6]

Strokes

Strokes occur in the brain when cell death results from lack of blood flow or bleeding.[7] A decrease in blood flow can occur because of a blood clot that forms locally or because of one that forms elsewhere in the body and travels to the brain. Blockage of a brain artery stops the flow of the glucose and oxygen that

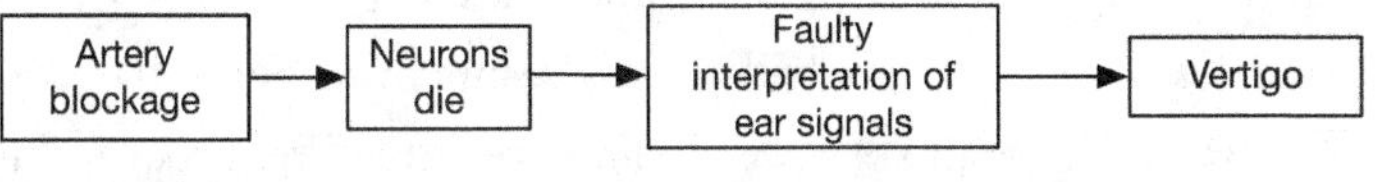

FIGURE 3.4 Vertigo arising from strokes. Rectangles indicate processes and arrows indicate causality.

provide energy to neurons and other brain cells. Neurons are particularly vulnerable because of their high demands for energy to fuel repeated firing. When the blood flow drops to less than 20 percent, damage to brain tissue becomes severe.

When neurons die, neural groups lose their ability to receive and send signals to other neural groups. For balance, the most serious strokes occur in the brainstem and the cerebellum since elimination of some of their neurons throws off the interpretation of information from the inner ear. Problems with ocular, motor, and cortical brain areas can also interfere with balance.

The most common disruption by which strokes can cause vertigo is shown in figure 3.4. People feel dizzy because the brain is falsely inferring that the head is moving when loss of neurons has led to a faulty interpretation of the signals that come from the inner ear. Again, balance disorders result from breakdowns in the normal mechanism.

Tumors

Brain tumors do not kill neurons, but they can also lead to vertigo through faulty processing of ear signals. For example, an acoustic neuroma is a benign, slow-growing tumor at the base of the brain that affects the auditory nerve, which also influences balance. As tumors grow, they compress nerves so that neural groups can no longer communicate with one another. As with vestibular neuritis, bad signals result from nerve damage. Tumors in the brainstem, cerebellum, or other areas that influence the interpretation of information from the inner ear can all lead to vertigo.

Multiple Sclerosis

A different kind of brain damage that can lead to vertigo occurs with multiple sclerosis, a disease in which the material that covers nerve cells becomes ineffective. Normally, myelin makes neural processing more efficient because it covers the axons that connect one neuron to another and makes the electrical signals move more than 10 times faster. With multiple sclerosis, the body's immune system disrupts the myelin, and the resulting ineffective neural signaling can lead to dizziness and vertigo.

Like vestibular neuritis and tumor-caused nerve compression, the vertigo that results from multiple sclerosis is an interference in the way that groups of neurons communicate with one another. Without such communication, neural groups cannot collectively collaborate to produce accurate interpretations of signals from the inner ear. Erroneous interpretation that the head is moving constantly generates the experience of vertigo.

Aging

A less extreme loss of myelin also occurs with age and helps to explain why memory is less effective in older people. Even without serious damage such as strokes and tumors, aging brains tend to suffer from a loss of neurons and synapses. These losses help to explain why dizziness and vertigo are far more common in the elderly than in younger people. In the United States, 24 percent of people older than seventy-two have dizziness, and falls are the leading cause of hospital admission and accidental death in older people. Some of these problems are due to general neural decline, but cases of inner ear dysfunction such as BPPV and Ménière's disease are also common.

Aging is the result of various developments: "The stability of posture and gaze during standing and walking is maintained by the rapid processing of vestibular, visual and somatosensory inputs in the central nervous systems, followed by outputs to the musculoskeletal and visual systems. Every factor in this system deteriorates during aging."[8] Inner ear disorders such as BPPV are more common in people over seventy because they have fewer of the hair cells crucial for balance and hearing.[9] The otoliths and semicircular canals also function less efficiently with age, although the prevalence of falls can be reduced by balance exercises such as tai chi.

Strokes, tumors, multiple sclerosis, and aging are just some of the ways in which brain damage can lead to vertigo. Epilepsy, head trauma, migraine headaches, Parkinson's disease, syphilis, Lyme disease, and meningitis can also contribute to vertigo because of breakdowns in brain areas that help to interpret inner ear signals. Heart disease can also disrupt the brain, as when blockage to a heart artery stops flow of blood to the brain and causes neural damage.

WORLD-SPINNING VERTIGO

My friend with stroke-induced vertigo feels like his head is spinning. The illusion of head spinning results from the erroneous inference that the head is constantly rotating. This inference can be caused either by bad signals coming from the inner ear or by bad interpretation of those signals in the brainstem and other areas. Besides this head-spinning vertigo, many people, including me, experience the world-spinning kind in which it seems that the head is stable but the room or outside location is moving. This inference is astonishing because in general the world does not spin round and round, so the inference that the room whirls goes against all of an afflicted person's previous experience.

To explain world-spinning vertigo, we need to consider how brains perceive motion. The mechanisms of motion perception explain why we are usually effective at recognizing moving objects such as dogs and cars, but also why we are occasionally subject to illusions. The experience that the world is whirling is a striking example of illusory motion perception that results from interactions between the balance system and parts of the brain that operate to recognize moving objects.

Motion Perception

Vision in humans and other animals is far more complicated than simply taking a picture with a camera that generates a digital image as a set of dots. Brains recognize lines, shapes, and objects, including their locations and movements. Such inferences require considerable processing of retinal signals by at least five brain areas.[10] Figure 3.5 shows the main brain areas used in visual processing, starting with V1 at the back of the head, which gets information from the eyes via the thalamus. V1 has millions of neurons that are tuned to discriminate between small changes in colors and spatial orientations. V1 feeds the results of its computations forward to V2, where neural groups carry out more

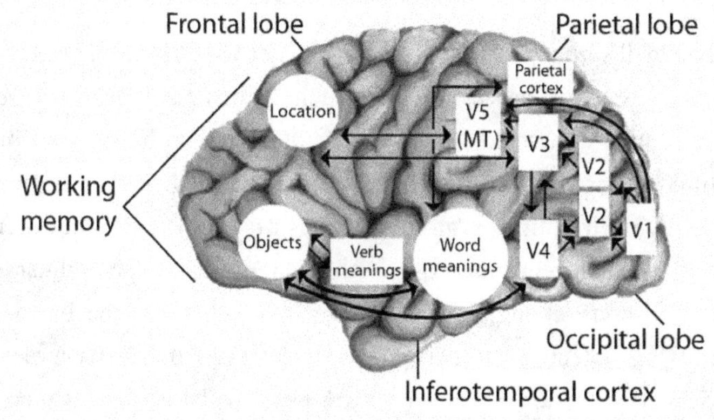

FIGURE 3.5 Organization of visual brain areas.

Reprinted with permission from Palma J. Longo, O. R. Anderson, and Paul Wicht, "Visual Thinking Networking Promotes Problem Solving Achievement for 9th Grade Earth Science Students," *Electronic Journal for Research in Science and Mathematics Education* 7, no. 1 (2002), https://ejrsme.icrsme.com/issue/view/732. Copyright © 2002 Association for the Advancement of Computing in Education (AACE), http://www.aace.org.

complex pattern recognition and also provide feedback to V1. V3 and V4 have connections from V1 and V2 and carry out more complex recognition of shapes and colors.

V5 is also known as MT, for "middle-temporal," and gets inputs from V1, V2, and V3 that contribute to the perception of motion. It contains neurons that are tuned to the speed and direction of moving objects. Crude information about motion comes from the retina when a light-sensitive cell goes from firing to not firing if an object passes by. V5 combines this information with object recognition to infer that a perceived object such as a car is moving. For example, if a white car is perceived to be in front of a brown tree, then not in front of the tree, V5 can infer that the car is moving.

Illusory Motion

Normally, the visual system works remarkably well, but the complexity of processing so much information from a rich and changing world sometimes leads to errors. More than one hundred visual illusions have been identified, many of them involving motion.[11] Watching videos on televisions, computers, and movie screens all requires the generation of visual illusions where mere changes in pixels are interpreted by the brain as moving objects such as people running. You might have experienced a motion illusion on a train or airplane when the train or plane beside you starts to move forward and you feel that you are moving backward. Another common motion illusion happens with wagon wheels or airplane propellers that are rapidly spinning in one direction but seem to be slowly spinning in the other direction.

Oscillopsia is a medically interesting illusory motion where people see the world as unstable and moving back and forth. It usually results from abnormal eye movements or an impaired vestibular-optical reflex. Oscillopsia is not the same as vertigo because it does not involve feelings of spinning or whirling, just the world jumping around. The normal process is as follows: "Humans are constantly moving and this self-motion causes the visual representations of these stationary objects to move across our retinas. If visual perception were based solely on the forward processing of sensory information, the position of objects in the world would appear to shift with each eye movement. In reality, our perception is that objects in the world remain stationary. This reflects the fact that we use an internal representation of the visual world that is actively constructed and spatially updated to account for our eye movements."[12] *However*, with oscillopsia this updating fails and generates the illusion that the world is jumping around. A similar operation can explain the origins of world-spinning vertigo.

Why the Room Spins

Because V5 is the main area of the brain for recognizing motion, I conjecture that it also contributes to the illusory perception that the environment is spinning. The contribution is not the normal one in which retinal changes tell the brain that objects are moving. No information comes from the retina saying that the room is moving. Instead, the moving room must be an illusory perceptual inference.

Adjacent to V5 is the medial superior temporal area (MST, or V5a), which also contributes to motion perception.[13] Moreover, MST receives signals not only from V5 but also from the vestibular system. I conjecture that MST, in cooperation with other brain areas, is capable of making inferences such as the following:

1. The vestibular system says that the head is constantly rotating.
2. But the neck muscles reporting through the spinal cord say that the head is not moving at all.
3. The best way to make sense of this contradiction is with the interpretation that the room is moving rather than the head.
4. Therefore, the room is moving.

This conjecture makes the testable prediction that brain scans of people suffering from world-spinning vertigo show more activation in MST than brain scans of people whose head-spinning vertigo results only from misinformation in the balance system.

For theories of perception that emphasize a direct relation between the world and what is perceived, this account sounds too cognitive. Is it legitimate to describe perception as making sense in a way similar to how scientists use hypotheses to makes sense of data? Yes, according to leading perception theorist Richard Gregory. Direct perception accounts have trouble explaining why so many visual illusions occur, but in Gregory's view such illusions are analogous to the false hypotheses that inevitably arise in science. Whether nonhuman animals experience vertigo is impossible to tell, but they seem to experience dizziness, as you can verify by doing a web search for dizzy cat, dog, and turkey videos.

NAUSEA

Many people with kinds of vertigo such as BPPV and Ménière's disease suffer from nausea, a feeling of stomach unease that often includes an urge to vomit. The word "nausea" comes from an ancient Greek word for seasickness with the same the root as "nautical." What brain mechanisms connect dizziness and vertigo with feeling nauseous?

Despite its etymology, many kinds of nausea have nothing to do with motion or balance. Other causes of nausea include food poisoning, excessive alcohol, pregnancy, chemotherapy, and disgust. I feel slightly nauseous when I see a dog walker stoop to pick up excrement with a plastic bag. It is easy to find an evolutionary explanation for a tendency to vomit after ingesting toxic food, since expelling the toxin should prevent further harm. Much more puzzling is why vertigo should be associated with a tendency to vomit.

A clue comes from the fact that the vestibular nuclei in the brainstem that control balance are close to a network of nuclei that control vomiting.[14] The latter nuclei have extensive neural connections with internal organs, including the stomach. Vomiting reflexes have ancient origins in animals that reject toxins, but motion sickness originated less than 10,000 years ago when humans invented boats; people do not get motion sickness from walking or riding horses.

The received theory of motion sickness is that it happens when unusual kinds of motion produced by boats, cars, or airplanes generate mismatches among different kinds of sensory inputs. One type of mismatch occurs between information from the inner ear and information from the eyes, as when reading in a moving car. Another type of mismatch occurs between information from the ears' semicircular canals and information from the otoliths. In both cases, the brainstem cannot make sense of the incoherent information, resulting in dizziness. Motion sickness occurs in fish, amphibians, and other animals with the same balance system as humans. Vertigo is also a mismatch between inner ear signals about head motion and other indicators that the head is stable.

Because of their proximity in the brainstem, the firing of neurons in the vestibular nuclei leads to firing of neurons in the nuclei relevant to vomiting. Some biologists claim that the brain perceives sensory conflict as a toxin for which vomiting is an appropriate response. But a more plausible hypothesis is that the vestibular-vomiting connection is just a side effect of the organization and interconnectivity of the brainstem. Evolution was not prepared for the strange signals that the brain receives from moving vehicles. Watching waves or herds of wild animals is not as disorienting as traveling in a boat or car.

Vertigo from BPPV and Ménière's disease is presumably much more ancient than motion sickness, but its connection with vomiting could also result from cross talk between vestibular nuclei and nuclei that affect vomiting. Such interactions need have no special function and would not necessarily be subject to evolutionary pressures because most kinds of vertigo occur in people well after their main reproductive ages. The connection between balance and nausea is accidental but still unpleasant.

It remains a mystery how the detection of sensory mismatches triggers activity in the vomit system, but here is one possibility. Chapter 4 describes a plausible mechanism that contributes to consciousness when neural representations (patterns of firing) evaluated as important are broadcast to other brain areas. For example, if you slice your finger and feel pain, the pain becomes conscious because nerves convey skin damage to your brainstem, which deems it important enough to broadcast it to the thalamus and cortex, where consciousness ensues. Similarly, sensory mismatches in motion sickness and vertigo are sufficiently unusual and important to be broadcast to connected areas that include the brainstem nuclei that control vomiting. Once stimulated, these nuclei send signals to the stomach and diaphragm that may initiate vomiting. The vestibular-vomiting connection is far from arbitrary because the vestibular nuclei already receive signals from the gut. The result is that the reflex of vomiting after eating bad food is generalized to produce vomiting after experiencing mixed-up balance sensations. When things don't make sense, prepare to throw up.

Symptoms of nausea illustrate how medical problems can arise not just from breakdowns in parts and in interactions, but also from interactions among interactions. The nausea that accompanies vertigo and motion sickness is a nonproductive side effect of faulty interactions between the vestibular neural network and the vomiting neural network.

FALLS

Falling is not a disease, but it can be a symptom of other diseases, such as the disorders that cause vertigo. When I had BPPV, I had to hold on to a wall, railing, or cane to keep from collapsing. Falls are an enormous medical problem, especially in older people, who can end up in nursing homes because of fractured hips. People with different balance disorders are prone to falls in different directions, such as forward-backward with BPPV and straight down with Ménière's disease. Parkinson's disease can cause postural instability and falls because of rigidity, slow movement, and impaired body perception.

Chapter 2 described the reflexes that maintain posture. The inner ear detects unusual motion that indicates falling, and it then signals the vestibular nuclei, which initiate movements in the neck and limbs via the cerebellum and motion areas. These movements should change the body's orientation and restore balance.

This postural mechanism can fail because of breakdowns in parts and interactions. If the inner ear is compromised because the canals or otoliths are not working properly, the vestibular nuclei receive misinformation that prevents them from sending useful signals about how to move the body to restore balance. Moreover, even good signals from the inner ear about possible falls can fail to initiate compensating movements if strokes or tumors damage the vestibular nuclei, other parts of the brainstem, or other brain areas such as the cerebellum.

Falling results when the brain and body are unable to satisfy all the constraints relevant to interpreting sensory inputs and adjusting the body to them. When balance is lost, the body undergoes a literal tipping point that inspires the metaphorical tipping points discussed in chapter 6.

Fall prevention techniques include changing environments by removing hazards such as loose rugs that are easy to trip over. Malfunctioning inner ears can be helped by fixing the damaged parts, such as by using the Epley maneuver to get crystals out of the canals and by using medications to reduce excessive fluid in the canals from Ménière's disease. Tumors can sometimes be removed from the brain to improve communication among brain areas, but not much can be done to fix damage from strokes.

To prevent falls by improving balance generally, people use balance exercises such as standing on one leg and shifting weight from one leg to the other. More systematic forms of exercise that improve balance include tai chi and yoga. All these balance exercises strengthen neural connections within and among brain areas so that posture can be maintained more efficiently. The simplest associations are learned when two neurons that are connected by synapses fire together, which increases the strength of the synapse and makes it more likely that they will fire together in the future. For example, when you see a new green avocado, the connections are strengthened between the neurons for green and the ones for avocado. Similarly, balance exercises make neurons for interpreting sensor signals more attuned to one another. People can unconsciously learn that if they move their arms and legs in particular ways, they can recover from loss of balance, thanks to an association between the proprioceptive sense of limb movement and the vestibular sense of losing and regaining balance.

Balance exercises probably also activate reinforcement learning in that success in carrying out a difficult task strengthens the neural connections that accomplished it. For example, if you succeed in balancing on one leg for thirty seconds, the neural connections that enable you to do it are rewarded through dopamine reinforcement and thereby strengthened.[15] Association and reinforcement learning explain the effectiveness of balance exercises without invoking bogus metaphors about tai chi and yoga, which are critiqued in chapter 7.

INDIVIDUAL DIFFERENCES

Why are some people so much better at balancing than others? Having BPPV or Ménière's disease marks individuals as having below-normal balance capability, and the incidence of most balance problems besides Ménière's climbs markedly with age. On the other hand, some people have extraordinarily good balance, such as the Wallenda family of tightrope walkers. Other great balancers include basketball players, ballet dancers, figure skaters, gymnasts, and steeplejacks who work on tall buildings.

Like individual differences in intelligence, balance variation results from a combination of genetics, epigenetics, and learning.[16] Perhaps the members of the Wallenda family have inherited genes for better balance, although they also had great learning opportunities from growing up surrounded by tightrope walkers. Not much is known about the genetics of vestibular disorders, although evidence is mounting that some cases of Ménière's disease are linked to family inheritance involving identifiable genes.

A second source of individual differences is epigenetics, which concerns chemical attachments to genes that determine whether the genes are activated or suppressed in their roles of producing proteins that affect behavior. Early reports find epigenetic changes as potential causes of both BPPV and Ménière's disease and as possible effects of treatments. Epigenetics challenges the traditional nature/nurture division because it shows how environmental influences can change how genes function.

Learning also contributes to good balance, as I mentioned with respect to the balance exercises that can help people protect themselves from falling. Association and reinforcement learning also helps people to become more proficient at the balance tasks required for tightrope walking, sports, and dancing. Practicing the challenging kinds of balancing required for gymnastics and figure skating strengthens the neural connections that enable numerous interacting brain areas to detect and modify the body's posture and motions. Moreover, balancing can be improved by learning specific techniques, such as the techniques by which figure skaters and ballet dancers spin rapidly without getting dizzy. Fixing their gaze on a single external object such as a door informs their brains that the world is not spinning. The eyes compensate for the disruption in the inner ear canals produced by the unusual exercise.

The interactions of genetics, epigenetics, and learning are shown in figure 3.6, which depicts how people's characteristics and capabilities are not simple matters of being "born this way" or being a "blank slate" filled in by learning and personal choices. Rather, genetic inheritance is modified by epigenetic change,

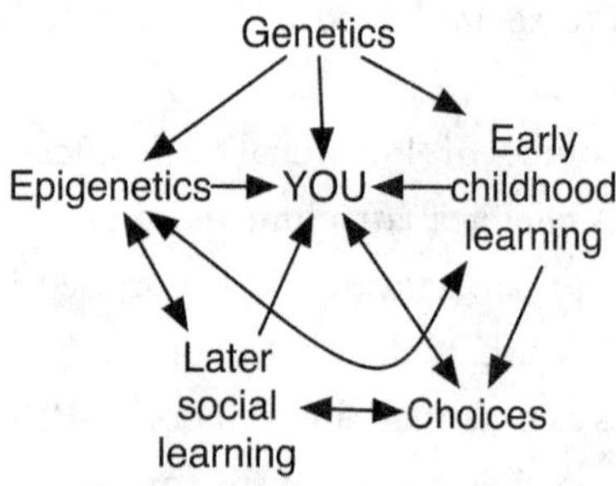

FIGURE 3.6 Interactions of genetics, epigenetics, and learning. Arrows show causal influences.

which can be affected by environmental influences, including the chemical state of the womb. Learning and choices help to select environments that modify epigenetics, leading to complex interactions that shape personality and identity. Good balancers such as the Wallendas, athletes, and dancers are neither just born nor made, but are rather shaped by the interplay of biological and social forces.

BALANCE BREAKDOWNS

People are healthy when the mechanisms that support functions such as metabolism, respiration, digestion, and locomotion are all working, with connected parts interacting to produce results that enable people to survive and accomplish their goals. Diseases arise from breakdowns in mechanisms caused by defective parts, connections, and interactions. Vertigo is not a disease but rather a symptom of a variety of diseases that involve different kinds of broken mechanisms.

The two main kinds of breakdowns that lead to vertigo occur in the inner ear and in the brain. Healthy balance depends on the proper functioning of the organs in each ear: the three semicircular canals and the two otoliths that send neural signals to the vestibular nuclei in the brainstem. These organs can be disturbed in various ways that produce vertigo, such as through the production of excess fluid (Ménière's disease, labyrinthitis) and through the movement of calcium crystals from the otoliths to the canals (BPPV). Treatment of these inner ear disorders requires fixing the underlying mechanisms, such as by moving the crystals out of the canals.

Brain breakdowns are harder to localize because they can occur in various areas that are important for computing balance, including the brainstem and cerebellum. Brain areas can cease to function properly because of damage to neural groups caused by trauma, strokes, or tumors. Then even normal signals coming from the inner ears can be misinterpreted to generate the experience of spinning or whirling.

Chapter 2 described four ways in which a toaster can break: through problems with parts, connections, interactions among parts, and interactions among interactions. Inner ear malfunctions and brain areas damaged by stroke illustrate part breakdowns, and damage to the vestibular nerve by tumors is a good example of connection breakdown. Brain damage can cause faulty interactions among the areas needed to coordinate balance. The problems in coordinating the brainstem areas that control balance and vomiting show the most complicated kind of breakdown: interactions among interactions.

These four kinds of mechanism breakdowns provide the causes of balance disorders such as vertigo. It is relatively easy to explain why a person's head or body seems to be rotating or spinning. When functioning normally, the inner ear organs detect motion of the head, so breakdowns generate incessant signals that the head is continuously rotating and seems to be spinning. Similarly, brain damage can generate signals of constant rotation.

More puzzling is how to explain the experience that the whole room is spinning, which requires attention to brain systems that perceive motion, including illusory motion. In room-spinning vertigo, neural groups in MT and MST mistakenly infer that the room is moving round and round. My speculation is that this inference results from a clash between (1) the information from ear or brain breakdowns that the head is rotating and (2) the observation from neck and body sensors that the head is not moving. The inference that the room is spinning resolves the impasse.

The balance system is like a group of scientists trying to make sense of all the data that they have collected. The system gets input from the ears, eyes, and other body parts and tries to generate a coherent interpretation that explains them, just as scientists try to construct theories that explain all the evidence they have collected through experiments and systematic observations. Often scientists succeed in coming up with successful theories such as genetics and relativity theory, but sometimes they make mistakes and get stuck with a bad theory, such as Ptolemy's view that the earth is the center of the universe. Similarly, the balance system sometimes makes mistakes in interpreting its bodily data and comes up with wrong interpretations, such as the spinning and nausea in vertigo. When sensemaking fails, people suffer.

The spinning room or spinning head is not just a conclusion but an experience that a person feels, along with conscious emotions such as surprise and distress. So a full explanation of vertigo requires a theory of consciousness.

4

CONSCIOUSNESS

In one of my favorite songs, Bob Dylan wails: "How does it feel, to be without a home, like a complete unknown, like a rolling stone?" The corresponding questions for balance are: how does it feel to be dizzy, nauseous, or have your head or the world spinning around? Even more interesting are the explanatory questions: why do people have feelings such as dizziness, nausea, and vertigo? These conscious experiences are usually accompanied by emotional feelings such as surprise, anxiety, and distress.

Aside from my explanation of room-spinning vertigo, my discussion so far has been a streamlined version of standard scientific accounts of balance. These accounts ignore the role of consciousness in balance and imbalance, which can be explained by synthesizing and extending current theories of how the brain generates feelings. Consciousness emerges from the same brain processes of sensemaking that produce balance and imbalance.

HOW DOES IT FEEL?

The balance system is a wonderful test bed for theories of consciousness because you are unaware of it when it keeps your body and vision stable, but you become conscious of it when you feel wobbly, are sick to your stomach, or feel you are whirling. What brings about the shift from unconscious but effective balancing to disturbing awareness of imbalance? An adequate theory of consciousness should be able to explain why walking and standing are usually done with little awareness whereas vertigo and falling are intense experiences.

Current neural theories of consciousness have not addressed balance and its disorders. I will describe how three of the most prominent theories can be extended to apply to balance and vertigo and discuss their strengths and weaknesses. One theory proposes that the major mechanism for consciousness is information integration, another that it is broadcasting across brain areas, and another that it is competition among representations. I will show how these mechanisms can be combined into a unified picture that explains why conscious imbalance differs from conscious balance. I will also consider alternative, non-neural theories, such as the idea that consciousness is a property of a nonphysical mind or soul, that consciousness is a property of everything in the universe, and that consciousness does not exist.

No definition of consciousness is generally accepted, but strict definitions that provide necessary and sufficient conditions are rare outside of mathematics. A more psychologically plausible way of analyzing concepts is to identify three aspects: exemplars, typical features, and explanations.[1] Exemplars are standard examples, including dizziness and vertigo as well as more common conscious experiences of colors, shapes, sounds, pain, emotions, and thoughts. The typical features of consciousness are experiences, awareness, attention, shifts between experiences, unity of experiences, and starts and stops of conscious awareness.

Imbalance displays the typical features of consciousness because dizziness and vertigo are experienced, people are aware of them, attention shifts to and from them, they start and stop, and they display unity when the experience of a spinning room integrates the room with the spin. Finally, consciousness has a valuable explanatory contribution when it makes sense of people's behaviors, experiences, and reports about their experiences. Balance concepts such as vertigo are explanatory in that they explain why people sometimes have difficulty walking and report being dizzy and falling. With these exemplars, typical features, and explanations, we can understand the meaning of consciousness without having a strict definition.

A good theory of why balance phenomena are sometimes conscious should be able to answer the following five balance questions:

1. Why are the operations of the balance system mostly unconscious and inaccessible?
2. Why do people suddenly become consciously aware of balance problems such as falling, dizziness, and vertigo?
3. Why are there qualitative differences between imbalance phenomena such as falling, unsteadiness, dizziness, nausea, and vertigo?
4. How are conscious experiences such as dizziness and vertigo sometimes combined with other mental experiences such as perception, pain, and emotions?
5. How can consciousness of perceptual experiences such as dizziness be elevated into consciousness of thoughts such as "I feel dizzy"?

All of these questions can be answered using a theory that unifies integration, broadcasting, and competition among neural representations.

INFORMATION INTEGRATION THEORY

Francis Crick made it respectable for biologists to probe the mysteries of consciousness, and his collaborator Christof Koch has continued the admirable adventure.[2] Koch's book *The Feeling of Life Itself* presents his newest take on consciousness with a defense of Giulio Tononi's theory that consciousness results from integrated information.

Information integration theory starts with five so-called axioms: that each conscious experience is (1) intrinsically "for itself," (2) structured into distinct sensory aspects, (3) informationally rich with abundant detail, (4) integrated in not being reducible to its components, and (5) definite in having contents and spatiotemporal properties that exclude other experiences. The last four are not as self-evident as axioms are supposed to be, but they are plausible generalizations about various kinds of consciousness. In contrast, the first is obscure because it does not explain what "for itself" means, except for describing physical elements that specify "differences that make a difference" to themselves; this description is uninformative and does not explain what making a difference means.

Koch states that "consciousness is a fundamental property of any mechanism that has cause-effect power upon itself."[3] This statement is ridiculous because consciousness would then include any machine that uses a feedback mechanism, such as the float valve in a toilet. When the toilet flushes, a valve in the water tank refills it with water until a hollow float connected to a lever rises enough to close the valve. The toilet causes itself to stop refilling with water when the float rises high enough to shut off the valve.

Figure 4.1 simplifies the causal structure of this apparatus, showing the causal feedback loop involving the water level, the float, and the valve. When you flush a toilet, pushing down the

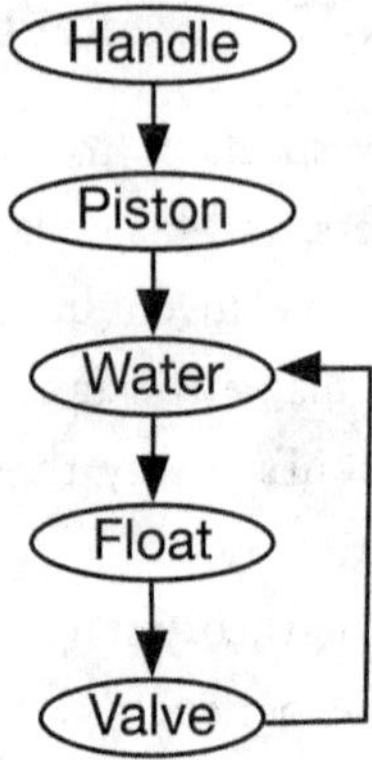

FIGURE 4.1 Feedback mechanism in a flush toilet. Arrows indicate causality.

handle raises the piston so the water flows into the bowl. Then the float lowers, opening the valve and putting water back into the tank. The toilet does contain some physical information; for example, the float can represent the water level and the valve can represent the water flow. You could even say that the apparatus integrates such information.

However, we have absolutely no reason to suppose that the toilet has even a little bit of consciousness, in contrast to humans and other animals. People are able to report their conscious experiences, and even fish have pain behaviors such as writhing that suggest that they might be conscious. In my book *Bots and Beasts*, I argue that consciousness first evolved with animals that have millions of neurons.[4]

Koch and Tononi may well bite the bullet and say that toilets are in fact a bit conscious, as they grant for bacteria and simple logic gates. Because none of these exhibit any signs of consciousness, a better strategy is to look elsewhere for a biologically plausible theory of the neural mechanisms that support consciousness. Instead, information integration theory offers a mathematical quantity called Φ (phi) that is supposed to measure the extent to which a causal mechanism cannot be reduced to its parts. Unfortunately, computing Φ requires considering all possible mechanisms that could operate in a system, a number that grows exponentially with the size of a system. For example, a neural group with only three neurons has just $2^3 = 8$ ways of combining them, but 100 neurons allow a larger number of possibilities that is 30 digits long, and calculating the number for billions of neurons would far exhaust the resources and history of the entire universe. Hence the Φ measure is mathematically useless for real systems. It also does not provide any explanation of why conscious experiences such as visual perceptions, pains, emotions, and thoughts are so different from one another.

Whereas Koch is extravagant in granting consciousness to bacteria and logic gates, he is stingy in insisting that computers cannot be conscious. His argument is based on two examples of currently successful technologies using deep learning networks and reinforcement learning. He says that these use feedforward neural networks without the causal feedback loops that he thinks are crucial for consciousness. But current computers such as smartphones are full of feedback loops, as when Google Maps readjusts your route when GPS shows that you missed your turn. Koch ought to conclude, based on information integration theory, that computers are already conscious. In contrast, I doubt that any computers are currently conscious, but it is an open question whether advances in computer hardware and software will eventually give computers the same causal powers as the brain to become conscious.

Superficially, information integration sounds like a good way to understand the balance system: chapter 2 described how effectively the brainstem combines inputs from the inner ear, eye, and other body parts. But the kind of neural information integration that I described as resulting from neural groups feeding into other neural groups is starkly different from the more abstract idea of Tononi. He says that integration occurs when mechanisms are not reducible to independent components, but he never gives a clear specification of reducibility or independence.

Information integration theory might explain the difference between unconscious balance and conscious imbalance by proclaiming a difference between the brainstem and the cortex, with the brainstem operating unconsciously while consciousness arises in the more integrated cortex. But the brainstem has feedback loops just like the cortex, so nothing suggests that it is any less integrated in Tononi's sense. Eminent neuroscientist Antonio Damasio argues that the brainstem is a major contributor to consciousness because of its role in mapping bodily states that generate feelings.[5]

Information integration theory is also incapable of answering other questions about balance, including the qualitative differences among various imbalance feelings, why such feelings often incorporate emotions, and how people become aware of higher thoughts such as "I feel dizzy." So we should not look to information integration theory to explain balance consciousness, but rather use the idea of information integration as performed by neural groups as part of the explanation of consciousness.

It might be objected that my understanding of information integration as a neural mechanism arbitrarily rules out the possibility of consciousness in nonbiological systems such as computers. Today our only plausible examples of conscious entities are biological systems, and the account of integration will need to expand if computers show signs of consciousness.

GLOBAL NEURONAL WORKSPACE THEORY

Stanislas Dehaene is a distinguished French cognitive neuroscientist who has done groundbreaking research on topics such as reading and numbers. In his book *Consciousness and the Brain* and numerous articles, he defends the theory that consciousness is global information broadcasting within the cortex that shares information throughout the brain.[6] Based on extensive brain-scanning experiments, he identifies four "signatures" of consciousness: (1) amplification of activity in parietal and prefrontal areas when a threshold for awareness is crossed, (2) occurrence of the P3 brain wave, (3) a sudden burst of high-frequency oscillations, and (4) synchronization of information exchanges across distant brain regions.

To explain these features, Dehaene proposes global neuronal workspace theory, according to which the brain contains a global workspace that allows many local brain areas to share information. Information becomes conscious when it accesses the workspace, which is not a particular brain area but is spread broadly across the cortex with major hubs in prefrontal areas. Then consciousness is just brainwide sharing of information.

Dehaene's theory has more clarity, biological plausibility, and computational feasibility than information integration theory but does not provide a good theory of balance consciousness. It has part of an answer to my first question about why balance is usually unconscious: activity in the balance system, including the vestibular nuclei, does not cross the relevant threshold and therefore does not broadcast to the rest of the brain. But global neuronal workspace theory has no mechanism for determining what neural activity crosses the relevant threshold to broadcast from the brainstem to the cortex.

Even more starkly, global neuronal workspace theory ignores the question of why particular experiences are associated with conscious events such as vertigo, dismissing as merely philosophical the experiential component of consciousness. It cannot explain why different conscious experiences have different feels and so cannot answer why dizziness, vertigo, and unsteadiness have similarities and differences. Moreover, without any neural bearers of conscious experience, global neuronal workspace theory cannot explain why imbalance experiences are combined with emotional reactions such as distress, let alone how people can be conscious of complete thoughts such as "I am dizzy." In sum, the theory valuably connects consciousness with signature phenomena such as brain waves, but it says too little about the nature of conscious experience. Broadcasting across brain areas is a promising mechanism for explaining some aspects of consciousness but needs to be unified with other mechanisms.

CONSCIOUSNESS AS SEMANTIC POINTER COMPETITION

My own theory of consciousness consists of three hypotheses that extend to cover balance.[7] These extensions answer the five balance questions at the beginning of the chapter about unconsciousness, becoming conscious, qualitative differences, integration with emotions, and self-representation. This theory can easily be broadened to include the most plausible aspects of information integration and neuronal broadcasting.

Hypothesis H1. Consciousness is a brain process resulting from neural mechanisms.

Chapters 2 and 3 described neural mechanisms for balance and vertigo, so it is natural to apply such mechanisms to explain their unconscious and conscious aspects. H1 has strong implications concerning what kinds of entities are capable of conscious experiences such as dizziness and vertigo. It rules out the occurrence of such phenomena in machines such as robots and in simple organisms such as bacteria that lack the neural mechanisms that produce balance and vertigo in people.

On the other hand, all vertebrates have vestibular systems similar to people, so they are candidates for having balance experiences. This possibility is tempered by the fact that the brains of fish and reptiles are much simpler than those of mammals, which have a more developed cerebral cortex, and those of birds, which have an analogous brain area called the nidopallium. I do not know whether fish and amphibians feel dizzy, even though their behavior indicates motion sickness. But mammals and birds have brains, inner ears, and behaviors sufficiently similar to humans that we can reasonably infer that they suffer from dizziness. If you doubt that animals get dizzy, do a web search for "dizzy cat" and "dizzy turkey."

Hypothesis H2. The crucial mechanisms for consciousness are representation by patterns of firing in neural groups, binding of these representations into semantic pointers, and competition among semantic pointers.

Chapter 2 explained basic ideas about neural representation. A single neuron cannot represent much, perhaps just that one of the hair cells in a semicircular canal is being stimulated. Each neuron contributes to the representational capacities of the brain to stand for things in the world, in this case fluid movement in the inner ear. A neuron can represent something in three ways: by firing or not firing; by firing fast or slow; or by firing with a particular pattern such as FIRE REST FIRE REST REST. All of these are ways of capturing what is going on in the inner ear that corresponds to what is happening in the world such as head rotation. Nevertheless, no evidence suggests that a single neuron can be conscious.

The representational capacity of brains explodes when neurons are combined into groups. Neurons form a group when they are tightly interconnected with one another through excitatory and inhibitory links that enable them to stimulate or discourage one another's firing. Groups of neurons are exponentially more powerful representationally than single neurons, so a group of neurons in the vestibular nuclei in the brainstem can collectively represent what is going on in the six semicircular canals in the inner ears. Still, the activity of such a group does not amount to consciousness.

Even more complex neural representations consist of groups of neurons whose firings are influenced by multiple neural groups. Chapter 2 described how neural groups that represent fluid motion in a semicircular canal can change the firing of neurons in a brainstem group that represents head rotation. This supergroup makes a complex inference about the motions in the fluids of the various canals. Then the pattern of firing in the supergroup represents the relevant occurrences in the world: the head is rotating left, rotating right, or not moving. But the supergroup is still not sufficient to produce consciousness, which captures only a tiny part of the brain's activities.

Because the brainstem has no capacity for language, the neural group's firing pattern may not say anything explicitly about the head. The firing pattern then more simply just represents "not moving," "moving left," or "moving right." The conscious experience of the head spinning to the left needs to bind a representation of the head with a representation of moving left. Even more complicated than head-spinning vertigo, room-spinning vertigo needs to bind a perceptual representation of the room with a representation for moving.

How such binding works is explained by Chris Eliasmith's semantic pointer architecture, a general theory of how the brain works.[8] Semantic pointers are neural representations (patterns of firing) formed by binding sensory, motor, emotional, and/or verbal representations, which are all patterns of firing in neural groups. Binding occurs by operations that weave together different representations in a way that allows them to operate as entities while retaining parts of their sensorimotor histories. For example, the semantic pointer for the concept *cat* is a neural process that combines sensory information about what cats look, feel, and sound like. The thought that the cat is jumping is formed by combining the *cat* semantic pointer with the *jump* semantic pointer.

Similarly, the thought that the head is spinning requires combining semantic pointers for *head* and *spin* into *head is spinning.* The result is typically more perceptual than verbal, although people can transform it into a sentence when they inform others about their spinning heads. Still, we have no reason to expect the semantic pointer for *spinning head* to be conscious, as almost all the activities of the approximately 86 billion neurons in the brain remain unconscious.

Of the countless neural representations operating in the brain at one time, only a small subset become conscious. One of the founding events of cognitive psychology was the recognition by George Miller in 1956 that short-term memory is limited to about seven items—without chunking them together, people cannot think of many things at once because their attention is limited.[9] Attention is a psychological mechanism that only allows into consciousness selected representations of what is most important for an organism. The best current psychological theory of attention is that it results from interactive competition among representations. For example, if you suddenly see a dog chasing you, you will attend to it rather than to the other things that are on your mind because your mental representation of the dog outcompetes your other concerns.

The corresponding brain mechanism is mutual inhibition among neural representations. One neuron can inhibit another neuron from firing by means of synapses that send chemical signals to the second neuron that reduce its firing rate. One neural group can inhibit the firing of neurons in the other group through inhibitory links between individual neurons. Similarly, semantic pointers can compete by inhibiting one another through having their individual neurons inhibit the firing of neurons in relevant groups. Then the neurons that represent *dog chasing me* fire strongly and inhibit the neurons representing other thoughts such as *I am hungry.* In the same way, the feeling of dizziness outcompetes more usual thoughts.

Balance experiences such as vertigo result from tiers of competition. In the vestibular nuclei, simple representations capture basic occurrences such as movement in the fluid in the inner ear, and these representations have to compete with alternative interpretations. At the next tier, groups of neurons integrate information from the inner ear with information from the eyes and other sensors to generate interpretations of head movement. Sensemaking ties together different kinds of sensory representations.

When the results are important to the organism, they compete to become conscious. One way that the brain makes evaluations of importance is via the emotional system such as through threat assessments in the amygdala. Dizziness and vertigo are threats because they are unpleasant and can lead to dangerous falls. The brainstem has ample connections with the amygdala and other areas that enable an emotional evaluation of imbalance interpretations. Firing of neural groups that connect vestibular representations with value representations helps the brain to decide what deserves attention.

Suppose that unusual fluid motion in the inner ears produces representations in the vestibular nuclei that repeatedly indicate *moving left*. That interpretation is sufficiently novel to warrant communication with the ocular and proprioceptive systems, and this communication yields the interpretation that the head is spinning or that the room is spinning. Either form of spinning is sufficiently disturbing to get a negative emotional boost and thereby compete with other neural events to become conscious. Vertigo is so novel and disturbing that it wins the attention contest and readily competes for consciousness, as do the dizziness and wobbliness that threaten to provoke falls. Hence imbalance happenings become conscious because the emotional assessment of their importance allows them to outcompete more mundane occurrences. In contrast, everyday balance events such as maintaining posture and eye movement while walking down the street are too routine to earn conscious attention.

Hypothesis H3. Qualitative experiences result from the competition won by semantic pointers that provide neural representations of sensory, motor, emotional, and verbal activity.

Representation, binding, and competition are crucial mechanisms for generating conscious experiences, but the nature of this experience depends on the neural firings that represent, get bound, and compete. For most words, the relation between the word and what it represents is arbitrary, as the English word "cat," French "chat," and German "katz" are not actually similar to cats in the world or in any kind of causal relationship to them. The rare exception for words is onomatopoeia, where there is some similarity between the word and the sound, as with "meow" for cat sounds. Semantic pointers, however, are neural representations—patterns of firing in groups—that can retain some of the characteristics of the sensory and motor modalities that generated them.

This property of *modal retention* explains how neural representations can retain some connection with the aspects of the world that incited them. Because of modal retention, different semantic pointers can generate different qualitative experiences because they originated with different interactions with the world. Hence semantic pointers for color experiences are different from ones for sounds or pains.

With balance phenomena, the relevant semantic pointers originated in the sensors in the inner ear, with semicircular canals and otoliths generating signals that directly represent fluid motion and indirectly indicate head motion. These signals then feed into neural groups that carry higher representations but still retain some of the sensory character of what initiated them, right through to the subsequent semantic pointer for *head is moving left.* The modal retention property of semantic pointers explains why the experience of blue is different from the experience of loud, and why dizziness feels different from vertigo. The qualitative experiences for vertigo, dizziness, and other imbalance phenomena are not in the ear sensors or in the neural representations of the vestibular nuclei, but rather in the brain processes that use semantic pointers to provide a more complex representation that retains some of the character of the initial sensory inputs.

Competition among neural representations to become conscious is another example of the process of parallel constraint satisfaction that I applied to balance in chapter 2 and to falling in chapter 3. The associations and inferential connections between representations serve as positive constraints that encourage them to be activated together, as when consciousness of the concept *brother* leads to consciousness of *sister*. But strong negative constraints also operate because of the limited capacity of consciousness, so inhibitory links between neurons prevent too many of them from firing out of control, as happens with epilepsy. The narrow focus of consciousness is the result of this competition of constraint satisfaction carried out by neural networks with excitatory and inhibitory links. Consciousness becomes coherent because constraint satisfaction organizes the competing demands of different inputs from the senses and memory, a splendid application of sensemaking.

Hypotheses H1–H3 are not directly testable by experiments individually, but they are worth considering because they work together to explain important phenomena about consciousness that evade other theories. The semantic pointer theory of consciousness has been used to explain qualitative aspects of consciousness, onset and cessation, shifts in experiences, differences across species, unity and diversity, and storage and retrieval. The theory's explanatory power is extended by considering balance consciousness.

Where in the Brain Is Consciousness?

The two prominent theories of consciousness I began with disagree about where in the brain consciousness operates. According to information integration theory, the neural correlates of consciousness are in the back of the brain in the posterior cortex, which is important for sensory operations such as vision. In contrast, the global neuronal workspace theory locates consciousness in the prefrontal cortex, to which different kinds of consciousness are broadcast. An adversarial collaboration is under way to perform brain-scanning experiments to determine which theory is closer to the truth about consciousness in the brain.[10]

I suspect that both theories are wrong because consciousness operates with a network of networks that cover different brain areas in both the front and back of the brain, including the brainstem. The University of Michigan's Department of Anesthesiology has a Center for Consciousness Science that uses the effects of anesthetics to study consciousness.[11] People undergoing surgery lose consciousness when they are given anesthetics, whose mechanisms are only beginning to be understood. The Michigan group has identified two networks independently identified by brain scans whose interactions are affected by anesthetics.

The default network gets its name because it seems to operate when the brain is not involved in specific activities. It consists of areas in the parietal, prefrontal, and temporal cortices such as the ventromedial prefrontal cortex and the posterior cingulate cortex. Whereas this network is most active when the brain is not engaged in attention-demanding tasks, the dorsal attention network responds to tasks that demand focus. This network includes prefrontal areas of the cortex, the insular cortex, the supplementary motor area, and possibly the cerebellum. The dorsal attention network and the default network are anticorrelated in that if one is active then the other is not, where "active" refers to the amount of neural firing in the brain areas in the network.

According to the Michigan group, loss of consciousness resulting from anesthetics such as propofol is correlated with disruption of a circuit that manages transitions between the two networks. Then consciousness operates not in a particular brain area or even a network of brain areas, but rather in the interactions of two networks that range all over the brain, from the prefrontal cortex to the cerebellum. Similarly, recent theories maintain that intelligence, creativity, emotion, and aging decline are all best explained by the interactions of brain networks rather than by activity in specific brain areas. Pain is not a simple matter of signals going from damaged tissue to a pain region in the brain, but rather requires interaction among brain areas such as the brainstem, amygdala, thalamus, prefrontal cortex, and others.

The network-of-networks view of consciousness fits well with balance-related experiences. Responses to bodily motions in the inner ear generate neural firings in the brainstem that are also influenced by interactions with other brain areas, including the visual cortex, motor cortex, thalamus, hippocampus, and cerebral cortex. The search for the "neural correlates of consciousness" in specific brain areas is a residue of the 1980s, when the brain was commonly thought to be highly modular, with particular areas performing particular tasks, as in a car engine. Brain-scanning research exploded this view as task after task was found to be associated with interactions among multiple brain areas. Claims about the modularity of mind have been cast aside in favor of emphasis on the *connectome*, the extensive axonal connections among different brain areas that perform interactively.[12]

Accordingly, imbalance consciousness is not localized but is rather distributed throughout the brain areas, as the Michigan network-of-networks view suggests. The key mechanisms of representation, binding, and competition occur in all these areas. A map of vertigo consciousness would look something like figure 2.7 with interacting brain areas. Like other kinds of consciousness, the sense of imbalance emerges from escalating emergence in neurons, neural groups, brain areas, networks of brain areas, and networks of networks. Competition between representations by means of inhibitory connections operates at all levels: between neurons, between neural groups, and between networks.

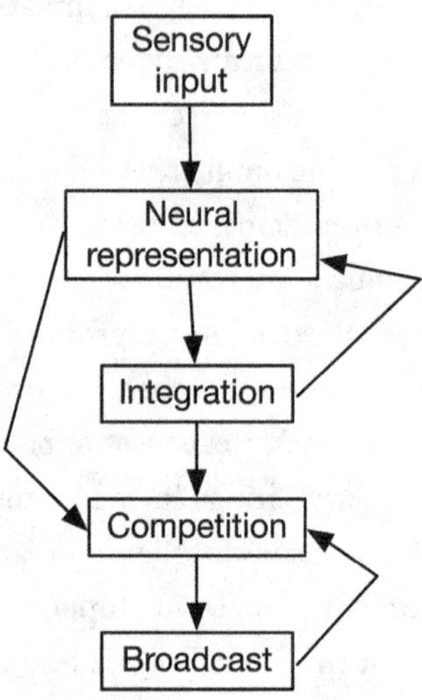

FIGURE 4.2 Unified picture of consciousness. Arrows indicate causality.

Figure 4.2 shows how to tie together the key ideas of the three theories of consciousness based on semantic pointer competition, information integration, and global broadcasting. The origin of conscious experience is in sensory inputs from familiar organs such as the eyes and also from the inner ear canals and otoliths. The senses generate neural patterns of firing in the brainstem, thalamus, and other areas. The resulting neural representations can be integrated and transformed into more complex representations such as interpretations of objects and movements. Both simple and complex representations compete with one another for the brain's limited capacity for attention, and winning the competition enables a representation to be broadcast to other areas of the brain by means of the axons that run back and forth among many parts of the brain. Communication to new brain areas sets off new competitions, as when the dizzy signal that originates in the brainstem arrives in the cortex and has to compete with higher thoughts expressible in language.

In this picture, information integration is an ongoing process in which neural groups feed into other neural groups with constant feedback, not an abstract and noncomputable mathematical quality. Broadcasting is not a special process but rather a side effect of neural groups carrying out evaluation, competition, and long-distance activation. The vestibular nuclei integrate information from the inner ears and other senses and broadcast the result to connected brain areas, but integration and broadcasting are effects of regular operations of neural groups. I now show that the resulting theory suffices to answer the five questions posed at the beginning of this chapter.

CONSCIOUSNESS OF BALANCE AND IMBALANCE

The major mechanisms of my theory of consciousness—representation, binding, and competition—extend to balance, but that is not enough to demonstrate that they should be accepted. I now show that this theory is superior to alternatives in explaining a full range of phenomena.

Balance Is Mostly Unconscious

Most of the time, you do not notice your balance system, despite the great contribution that it makes to your movement, stability, and vision. Walking down the street and looking around are routine thanks to the operations of the vestibular nuclei and the reflexes that keep your body and eyes steady. Consciousness does not apply to these mechanisms for the same reason that it does not apply to most of what goes on in your body and brain. For example, you have no awareness of digestion until something goes wrong and you end up with pain or bloating. The routine physiological operations of digestion and balance are not important enough to the organism to win the competition for becoming consciousness. Special circumstances such as high-wire tightrope walking stimulate emotions such as fear that can bring usually mundane aspects of balance to consciousness.

Why is consciousness so limited? The answer to this question depends on the functions of consciousness that are not obvious. No definitive view explains why animals evolved to be conscious, but plausible functions include contributions to learning, action, and communication. Your brain has billions of neurons operating all at once, but this parallel or simultaneous operation contrasts with the serial operation of your body. You cannot run in two directions at the same time, so your brain needs to narrow down action to a small number of actions that it can perform simultaneously. Thinking is parallel but action is serial, so consciousness makes your brain focus on what to do next. Consciousness also helps you to deal with other people, because you can teach them how to do useful things like make food by becoming aware of how you make food. But neither focusing nor teaching is required for mundane balance operations that keep you upright, moving, and seeing the environment. So balance usually remains unconscious.

Balance Becomes Conscious When Disturbed

However, when things go wrong, the conscious brain is ready to become aware of them. Dizziness, vertigo, and unsteadiness are all threats to the integrity of the body because they often lead to falls. All demand focus and may be worth communicating to other people who can help against falling or provide sympathy. Accordingly, the mechanism of competition favors balance processes that are unusual and important.

Emotion is a major contributor to the evaluations that determine what neural representations break through into consciousness. The connections from the vestibular nuclei and other balance-affected brain areas to emotion-relevant areas such as the amygdala ensure that disruptions in the normal balance operations can be marked as serious and therefore get increased activation through feedback loops from emotion to balance.

Qualitative Differences

Such loops explain why imbalance phenomena become conscious but not why they generate particular feelings. Why are dizziness, wobbliness, and vertigo different feelings? In 2009, the Committee for the Classification of Vestibular Disorders produced a comprehensive classification of symptoms that provides a guide to the conscious experiences that need to be explained by a theory of balance.[13] The committee carefully distinguished between varieties of vertigo, dizziness, and unsteadiness that I will connect with neural mechanisms. Whereas philosophers sometimes speak vaguely about a conscious experience using phrases such as "there is something that it's like," I will consider specific experiences and show why they fall under a common set of explanations and also why they have differences that can be explained.

To distinguish dizziness from vertigo, the committee characterized dizziness as the "sensation of disturbed or impaired spatial orientation without [a] false or distorted sense of motion."[14] In contrast, vertigo does have a distorted sense of motion; this sense includes spinning but can also include other false motions such as swaying, tilting, bobbing, bouncing, or sliding. The committee distinguished the "internal" sensation of self-motion from the "external" sensation that the visual surrounding is spinning or flowing. This internal/external distinction is the same as the distinction in chapter 3 between head-spinning and world-spinning vertigo. Under the general category of postural symptoms, the committee included falls and unsteadiness, defined as "the feeling of being unstable while seated, standing, or walking without a particular directional preference."[15] References to feelings and sensations indicate that these are all conscious experiences.

Why do people experience the disturbed spatial orientation that marks dizziness? Spatial orientation is perceived by several modalities, including vision, which records where your body is in its surroundings; proprioception, which records the position and movement of joints and muscles; and the inner ear organs, which record the orientation and movement of the head. Each of these recordings is translated into neural representations that normally are coherent with one another and therefore can be bound into unified representations that capture the location of the body in space. With normal balance, the resulting representation remains unconscious.

As with vertigo described in chapter 3, the disturbance of dizziness can result from unusual fluid movement in the semicircular canals and the otoliths or from brain damage that bungles the interpretation of this fluid movement. When the disturbance threatens the well-being of the body, it becomes sufficiently important to break through into consciousness by outcompeting other experiences and thoughts. What it feels like to be dizzy is therefore the combination of the proprioceptive, visual, and balance sensory inputs with an interpretation that these are not coherent, meaning that something is wrong. The neural representation of dizziness is not a simple, localized pattern of firing like the sensation of redness, but rather the result of binding several kinds of sensory inputs, including representations of the body, into a semantic pointer. The feeling of dizziness is then a neural process that integrates signals from inner ears with representations of the body, conveying the message that the body is at risk of falling.

Vertigo results from the same mechanisms as dizziness, with the additional inference that something is inappropriately moving: the head or its surroundings are spinning or performing some other motion. In head-spinning vertigo, the visual and proprioceptive sensory representation of the head is bound with motor representations of movement. In room-spinning vertigo, the binding is instead between representations of the room and of movement.

Either way, the resulting representation is surprising and threatening, as determined by the amygdala and other areas for emotional evaluation such as the ventromedial prefrontal cortex. People therefore feel vertigo because of the mechanisms of neural representation, combinatorial binding, and competition among the resulting representations. Vertigo feels different from mere dizziness because the additional inference about spinning creates a richer semantic pointer.

Feelings of unsteadiness and falling require different bindings of visual, proprioceptive, and balance representations. Such feelings are unlike vertigo because they do not involve bindings of representations for motion of self or surroundings. They are also unlike dizziness because they concern not just disturbed spatial orientation but also the imminence of the body collapsing, which requires binding a representation of the body with representation of movement downward. Again, the resulting representation is sufficiently threatening to life goals that it outcompetes other representations and enters consciousness in the network of networks where conscious experiences occur.

So the neural mechanisms of representation, binding, and competition together explain the feelings of dizziness, vertigo, and unsteadiness that occur when balance goes awry. The differences between these occurrences result from differences in inputs from visual, bodily, and balance systems, with varying inputs leading to varying neural firings and alternative bindings producing different neural representations that produce different feelings.

Another conscious experience associated with the balance system is nausea, which was not included in the committee's classification of vestibular symptoms. But people with vertigo often experience nausea in a way that is explained by the same theory of consciousness. The neural theory of nausea in chapter 3 included interactions in the brainstem between vestibular nuclei and the vomit network, both of which are also connected to the stomach and diaphragm, whose muscle contractions can induce vomiting. The conscious experience of nausea results from the binding of representations that combine imbalance and the need to vomit, which explains why nausea feels different from dizziness, vertigo, and unsteadiness, which lack the visceral contribution. The resulting representations are sufficiently unusual and unpleasant to compete their way into consciousness. What it feels like to be nauseous is explained by the operations of the brain, stomach, and other body parts.

Integration with Emotion

My discussion of why dizziness, vertigo, and unsteadiness win the competition to become conscious introduced emotional evaluation as a contributing mechanism. Anyone who has had imbalance feelings knows that they are unpleasantly accompanied by emotions such as surprise, distress, anxiety, and even fear, all of which motivate people to seek help. An account of imbalance consciousness needs to explain how feelings such as vertigo get combined with emotional reactions.

The well-developed semantic pointer theory of emotions meshes perfectly with balance consciousness.[16] In brief, emotions are semantic pointers that bind (1) neural representations of a situation such as standing on a high tower, (2) neural representations of bodily processes such as heart rate, breathing rate, and stomach feelings, and (3) cognitive appraisals performed by neural computations of the relevance of the situation to a person's goals. For balance problems, the representation of the situation includes the balance information such as stability or instability. The composite balance information from the visual, bodily, and ear systems can then be bound with the emotional reaction that takes into account additional body information and the inferential process of evaluating relevance to the system goals. Hence reactions such as dizziness, vertigo, and unsteadiness bind with emotional reactions to produce semantic pointers that easily compete for consciousness, as when people feel distressed that they are dizzy.

The balance system can also be the source of positive emotions such as exhilaration and pride based on the performance of challenging tasks like tightrope walking, skating spins, basketball dunks, and dance pirouettes. Physical actions that require consummate balance generate positive emotions because they satisfy important goals, thereby elevating the usually mundane accomplishment of bodily adjustment into conscious recognition.

Integration with Language

No one needs language to feel dizzy, as is evident from the whirling but nonvocal cats, dogs, and turkeys you can view on the web.

Dizziness, vertigo, and unsteadiness are nonverbal feelings, but people translate them into words that can be used to describe these states to other people such as doctors. People then become conscious of linguistic representations such as "I am dizzy" and "The room is spinning." A theory of consciousness must apply to linguistic thoughts as well as to bodily sensations.

Information integration theory and global neuronal workspace theory say nothing about the processing of language. But extensive research shows how semantic pointers can represent complex sentences by binding concepts into representations that can be embedded in other representations.[17] The resulting neural representations of sentences can then be bound with emotional reactions as in "I am unhappy that I am dizzy" and can compete with uninteresting representations to become conscious. People can then think and talk about their imbalance experiences.

I have argued that balance-related conscious experiences are brain processes resulting from the mechanisms of neural representation, binding, and competition among representations. The conclusion is justified because these mechanisms provide the best available explanations of unconsciousness, becoming conscious, qualitative differences, and integration with emotion and language.

SOLVING THE HARD PROBLEM OF CONSCIOUSNESS

The so-called hard problem of consciousness is explaining how physical systems such as the brain can have experiences and feelings and why "there is something it is like" to have such experiences.[18] The semantic pointer theory offers a general solution that covers all kinds of consciousness, including perceptions, pains, emotions, and thoughts. Here I offer a solution for the particular kinds of conscious experiences that arise in balance disorders, including dizziness, vertigo, and unsteadiness. Similar solutions operate for other kinds but depend on the neural mechanisms that are specific to them.

For balance disorders, the hard problem is why people sometimes *feel* dizzy or unsteady and why they *experience* various kinds of vertigo. These feelings and experiences are not just behaviors, although they cause behaviors such as staggering, falling down, moaning, and reporting that something is wrong. People who think that the hard problem is insoluble claim that neuroscience can never say why people have such experiences because, no matter what mechanisms are specified, we can still imagine beings that have such mechanisms but lack consciousness.

This argument uses a thought experiment that merely expresses a philosophical prejudice while ignoring the logic of explanatory identities, which have been repeatedly successful in the history of science.[19] An explanatory identity occurs when a familiar occurrence such as heat or lightning is identified with a mechanism on the grounds that the identification provides the best explanation of the evidence about the occurrence. For example, we have good reason to believe that lightning is electrical discharge from clouds because that hypothesis explains observations such as flashes of light, destruction of objects struck by lightning, and the generation of thunder. We can easily imagine that there could be electrical discharge between clouds without lightning, but that flight of fancy is irrelevant to the evidence-based determination that, as a matter of fact, lightning *is* electrical discharge.

Similarly, the explanations I have given for how neural mechanisms explain vertigo are good grounds for inferring that the experience of vertigo is a brain process resulting from the mechanisms specified, including neural representation, integration, competition, and broadcasting. That experience is different from the firing and interaction of neurons, but explanatory identities often have emergent properties. Emergence occurs when a whole system has a property that is not a property of its parts because it results from the interactions of its parts.

Table 4.1 displays emergence in numerous explanatory identities. For example, we have abundant evidence that air is not an element as the ancient Greeks assumed. Rather, it is a mixture of gases consisting of colliding molecules with emergent properties such as forming hurricanes. No one gas molecule is a hurricane, but large numbers of gas molecules have emergent effects such as blowing trees down. A hurricane is not just the sum of the forces of individual molecules because the collisions of the molecules with one another also affects their motion, along with heat and water vapor. Similarly, consciousness is a property of networks of networks of groups of neurons even though no particular neuron, neural group, or network is by itself conscious.

The fourteen explanatory identities in table 4.1 show how a puzzling target can be explained by identifying a source that provides a mechanism. We can identify the source with the target because the interactions of parts have emergent effects that explain the evidence. Analogously, we should not be puzzled that complex neural mechanisms can have emergent effects such as the feelings of vertigo and dizziness. The explanatory gap between biology and experience is being filled in the same way that science has filled previous explanatory gaps such as between water and its constituent atoms. Vertigo and other kinds of balance-related consciousness are emergent properties of brain processes resulting from interactions among neural networks whose activity derives in part from sensory inputs.

TABLE 4.1 Emergent effects of common entities and processes

Target	Source	Parts	Interactions	Emergent effects
Air	Mixture of gases	Molecules	Collisions of molecules	Hurricane, respiration
Blood	Cells in liquid	Cells, plasma	Suspension of cells	Flowing, oxidation
Cloud	Mass of liquid droplets	Water droplets	Collision, combination	Shape, motion, precipitation
Electricity	Flow of electrons	Electrons	Repulsion	Current, sparks
Fire	Rapid oxidation	Molecules	Oxidation	Flame, heat
Gold	Element with 79 protons	Atoms, protons, electrons	Attraction	Solid
Heat	Transferred energy	Molecules	Generation, transfer	Temperature, melting
Light	Wave-particle	Quantum = wave/ particle	Interference	Illumination
Lightning	Atmospheric electricity	Clouds, electrons	Charges	Flashes, burns
Magnetism	Electrical attraction and repulsion	Electrons, fields	Repulsion	Attraction, repulsion
Salt	Sodium chloride	Atoms	Bonding	Taste
Star	Luminous gases	Atoms and plasma	Fusion	Light emission
Thunder	Sound caused by lightning	Water droplets, electrons	Heating	Booms, shock waves
Water	H_2O	Atoms	Bonding	Freezing, boiling

In sum, the hard problem of consciousness is being solved by three complementary strategies. First, research is revealing how different experiences result from different neural processes based on semantic pointer competition. Second, we can recognize that feelings are emergent properties of these processes. Third, reflection on the history of science shows that imagining how feelings might not be brain processes is as irrelevant as imagining that air might not be a mixture of gases.

NON-NEURAL THEORIES OF CONSCIOUSNESS

I have extended brain mechanisms for balance and vertigo into explanations for consciousness, but prominent views maintain that consciousness does not result from biological mechanisms. These views include dualism, which claims that mind and body are fundamentally separate, panpsychism, which claims that everything in the universe has a bit of consciousness to it, and denial, which says that consciousness is nonexistent.

Dualism

In the general population, the most common view of mind and consciousness is dualism, which says that mind and body are fundamentally different and consciousness is the property of mind rather than body.[20] Billions of people who adhere to religions such as Christianity, Islam, and Hinduism believe that consciousness continues after death, so it must be a property of a nonphysical soul rather than the brain, which no longer functions. I am not aware of any discussions of whether dizziness, vertigo, and nausea survive death, but it is easy to imagine them being afflictions in hell.

Dualism faces well-known problems. First, evidence is missing that anyone has ever managed to survive death, so that beliefs in souls and an afterlife are merely matters of faith, which is indistinguishable from wishful thinking. Second, dualism has never been able to explain how mind and body can interact if they are completely different substances. This problem arises for the balance system, as there is extensive evidence for the contributions of the inner ear, brainstem, cerebellum, and visual cortex to maintaining balance. How any of these could influence a nonphysical soul is a complete mystery.

Third, dualism offers no answers to the five balance questions about consciousness. If consciousness is nonphysical, why can't it become aware of basic operations of the balance system, such as fluid movements in the inner ear? Such a comprehensive consciousness would bely the fact that balancing is usually unconscious until unusual events suddenly make people dizzy, nauseous, or overwhelmed by vertigo. Equally unexplained is why dizziness, nausea, and vertigo are different feelings, as in the varieties of vertigo where some people report that their head is spinning and others that the world is spinning. Finally, dualism says nothing at all about how loss of balance can be combined with other feelings such as distress and with linguistic descriptions such as "I think I'm falling."

Philosophical defenses of dualism are usually based on thought experiments such as the one I criticized about imagining beings with all our neural mechanisms but without consciousness. Philosophical thought experiments are about as reliable a guide to reality as religious texts and Republican tweets. Dualism's scant plausibility derives from the historical absence of biological explanations of consciousness. My account of imbalance consciousness illustrates how this explanatory gap is rapidly closing.

Panpsychism

Another alternative to a neural theory of consciousness is panpsychism, which claims that everything in the universe has at least a bit of consciousness to it.[21] Whereas dualism typically claims that human minds have consciousness because God put it there, panpsychism offers a different explanation of how minds became conscious. It agrees with dualism that consciousness is too astonishing to have evolved biologically like vision and digestion, but maintains that consciousness belongs to everything that exists. Not just animals but also plants and rocks and even atoms have a tiny bit of consciousness. Bits of consciousness somehow add up to the full-blown consciousness that humans possess.

The most serious problem with panpsychism is the lack of evidence that anything besides animals possesses consciousness. We have good reason to believe that people are conscious because they report conscious experiences such as seeing, hearing, and pain. Moreover, our behavior fits well with us having these frequent conscious experiences, as when we moan from toothaches or duck things thrown at our heads. With nonhuman animals, we do not get the same verbal reports, but their behavior is consistent with conscious experiences such as pain, as when a cat that has banged its paw licks it to recover. Moreover, the brains of all mammals and birds are sufficiently similar to those of humans that they generate experiences such as dizziness. In contrast, trees, rocks, toilets, and atoms display absolutely no behaviors suggesting that they have any degree of conscious experience.

The panpsychist argument that everything must be conscious in order to explain how people are conscious has two flaws. First, nature is full of phenomena that display emergent properties that are properties of the whole but not of the parts that make them up. Table 4.1 provides examples, and others include flocks of birds that maneuver as a whole through the interactions of individuals, and avalanches, which can destroy forests even though individual snowflakes are inconsequential. Similarly, our brains have numerous emergent properties such as their ability to represent the world through interactions of millions of neurons, each of which has little effect by itself. Emergence shows that there does not need to be bits of consciousness in small parts for whole brains to be conscious.

The second flaw in panpsychism is that it says nothing about *how* the bits of consciousness that are supposed to operate in atoms, rocks, and plants add up to the full-fledged consciousness in human beings, with our perceptions, emotions, and thoughts. More specifically, panpsychism says nothing about balance-related consciousness. As far as we know, atoms, rocks, and plants don't get dizzy, so they don't have bits of dizziness that somehow add up to the dizzy experiences of human beings. If anything can be conscious, then the vestibular nuclei in the brainstem should also be capable of consciousness, but alone they do not yield any experiences such as what is going on in the inner ears. Such experiences emerge from interactions of many brain areas.

Panpsychism cannot even sketch an explanation of why most balance operations are unconscious and why we suddenly become aware of dizziness or vertigo. It has nothing to contribute to why experiences of dizziness, vertigo, and nausea are so different. Similarly, panpsychism does not begin to explain how these experiences can combine with emotions and linguistic descriptions. Therefore, panpsychism is useless as a theory of consciousness for balance and imbalance.

Substrate Independence

Another philosophical account of the relation between mind and matter is that mental processes are independent of any particular physical substrate and can operate in a wide range of substances, including brains, computers, and perhaps alien force fields. This account was plausible in the 1970s when the view of thought as digital computing was dominant, but it has become dubious since the rise of cognitive neuroscience in the 1980s. Human neural balance mechanisms based on the vestibular system are importantly different from the sensors and control systems that enable robots to balance.

Moreover, the general implausibility of substrate independence is shown by considering how energy operates in different substances, as sketched in this argument:[22]

1. Real-world information processing depends on energy.
2. Energy depends on material substrates.
3. Therefore, information processing depends on material substrates.
4. Therefore, substrate independence is false.

Brain balancing depends on chemical energy provided to cells by mitochondria, whereas robot balancing depends on electrical energy provided by batteries. Hence human balancing can be identified with neural mechanisms even though robots have their own kind of balancing.

Denial

A different strategy for dealing with the problem of consciousness is simply to deny that it exists. This strategy was employed by behaviorism, an influential twentieth-century approach to psychology and philosophy. Behaviorism rejected consideration of people's inner mental lives as unscientific or metaphysical. Instead, psychology should concentrate merely on looking for connections between environments and behavior without any internal mental connectors.

By the 1960s, behaviorism had clearly failed because it could not even explain simple behaviors such as rats running mazes and pigeons learning patterns. It was appropriately replaced by cognitive psychology, which recognized minds as information processors that use a variety of mental representations and problem-solving strategies. Cognitive psychology did not yield a general theory of consciousness but did not deny its existence. Since the 1980s, cognitive psychology has merged with neuroscience to yield powerful explanations of a huge range of mental capacities, and consciousness is being added to the list.

Behaviorism would obliterate talk of experiences such as dizziness, nausea, and whirling rooms reported by billions of people. No scientific reason rules out such occurrences as targets of explanation, and claims that science should be limited to what can be observed violate general recognition in the philosophy of science that theories valuably go beyond observation. Behaviorism cannot explain why people go to their physicians complaining of dizziness or why more than one hundred languages serviced by Google Translate have their own words for "dizzy."

Another strategy for denying consciousness is *eliminative materialism*, which claims that as science develops the concept of consciousness will simply drop out of consideration along with abandoned concepts such as ether, phlogiston, and vital force. Elimination is the right strategy for some concepts such as soul, immortality, and possibly free will. But the personal and medical importance of balance disorders such as dizziness and vertigo that include conscious experiences advises against premature elimination. The rightly preferred strategy in neuroscience today is to explain consciousness rather than to supersede it. Overall, therefore, denial is a feeble strategy for consciousness, and we should press forward to see how well feelings can be explained by the operations of the brain.

CONSCIOUS BALANCE

Balance is like breathing—we take it for granted unless it is not working as it should. Many people with COVID-19 could breathe automatically for decades before lung infections made breathing suddenly labored. Similarly, balance is unconsciously effective until problems such as dizziness and vertigo force people to think more deliberately about how to stand and move. Imbalance problems such as wobbliness and nausea come with feelings that make people aware that something is wrong with their bodies.

The conscious experiences generated by the balance system should not be denied or rendered mysterious by dualism or panpsychism. Instead, the prospects are increasingly great for describing neural mechanisms that maintain bodily balance and sometimes break down to produce conscious disturbances such as vertigo and unsteadiness. Three neural mechanisms that contribute to these experiences are representation by patterns of firing in groups of neurons, binding of these representations into more complex ones such as semantic pointers, and competition among representations for the scarce resources of consciousness. Consciousness results not just from the firing of neurons in specific brain areas, but also from interactions among interactions in networks of neural networks.

The limited range of consciousness might just be an accident of biological evolution, but several factors mark it as a feature rather than a bug. If people were conscious of more that is happening in their bodies, such as the balance reflexes, then they would lose the focusing function of consciousness for important behaviors like acting on threats and teaching children. Feelings direct the body to deal with what matters to its survival and needs. Moreover, consciousness probably comes with a high metabolic cost because the neural firings required for representations of representations across numerous brain areas require large amounts of energy, in animals that are already expending a large proportion of their energy supplies on brain operations. Restricting consciousness to matters most important is then part of the information-energy trade-off that all organisms must accommodate.

My semantic pointer competition theory explains why balance is usually unconscious but enters consciousness when problems arise. Unlike alternative theories based on information integration and the global neuronal workspace, my theory also explains why different imbalance experiences such as vertigo and nausea come with different feelings. Using the same neural mechanisms for sensemaking, the theory also explains how feelings of imbalance can combine with emotions and verbal expressions.

The mental significance of balance goes well beyond keeping the body stable. The next chapter describes how the balance system contributes metaphorically to all domains of human existence.

5

HOW METAPHORS WORK

Human thought often progresses from metaphors to mechanisms, such as from fuzzy ideas about spirits in plants and animals to mechanistic explanations based on genetics and metabolism. But metaphors remain important because they enable us to connect puzzling aspects of life with things we understand better. Metaphors are paradoxical because they require saying that something is what it is not, for example, that budgets balance. But such literal falsehoods open up new avenues to creativity, understanding, and elegant expression. The etymology of the word "metaphor" traces it to an ancient Greek word for "transfer" or "carrying across." Balance metaphors transfer meaning and emotional value from familiar actions like standing up straight to novel applications like budgets.

I have used neural mechanisms to give literal, scientific explanations of balance phenomena operating in the brains and bodies of humans and other vertebrates. But balance concepts flourish in other areas of human thought, including science (chemical equilibrium), medicine (balanced diet), psychology (stable personality), art (balanced composition), and philosophy (reflective equilibrium). I examine balance metaphors to show their huge impact on how people think about the world and to grasp how they work in human minds.

This understanding comes from mechanisms that explain the mental and neural processes by which people use metaphors for many purposes, from explanation to entertainment, all varieties of sensemaking. Metaphor elevates balance from mundane bodily concerns to the heights of human reflection on the meaning of life. My analysis draws on insights about the nature of metaphor from psychologists and philosophers, particularly Keith Holyoak, Mark Johnson, and Laurence Barsalou.

POETIC METAPHOR

Keith Holyoak's book *The Spider's Thread* is a rich and eloquent discussion of metaphor in mind, brain, and poetry.[1] The title is from the poem "A Noiseless Patient Spider" in which Walt Whitman describes his soul as flinging a gossamer thread that it hopes will catch somewhere. Holyoak's analysis of poetic metaphor provides a good starting point for understanding balance metaphors. He describes a metaphor as a comparison between a target to be understood and a source that illuminates it.

Here is the Whitman poem:

A noiseless patient spider,
I mark'd where on a little promontory it stood isolated,
Mark'd how to explore the vacant vast surrounding,
It launch'd forth filament, filament, filament, out of itself,
Ever unreeling them, ever tirelessly speeding them.
And you O my soul where you stand,
Surrounded, detached, in measureless oceans of space,

Ceaselessly musing, venturing, throwing, seeking the spheres to connect them,
Till the bridge you will need be form'd, till the ductile anchor hold,
Till the gossamer thread you fling catch somewhere, O my soul.

In this poem, the target is Whitman's soul, which seeks to understand its boundless world. The metaphorical source is the spider, which persistently launches filaments to latch on to something, suggesting how the soul's efforts might eventually connect with the world. The spider source highlights the venturesome nature of the target soul, and the comparison increases our understanding of the soul's quest and provides enjoyment through Whitman's evocative language.

Similarly, in the economic metaphor *balanced budget*, the target is the process of coming up with a budget where expenditures equal income, just as in the source, a balance scale, the weight on the left pan of the scale should equal the weight on the right pan. In the psychological metaphor *steady person*, the target personality is illuminated by comparison with objects and bodies that are unbalanced.

Holyoak contrasts brief, focal metaphors that merely combine a pair of concepts such as *steady* and *person* with extended metaphors that do a systematic mapping between the source and target. Extended metaphors provide an analogical resonance that transfers important aspects of the source over to the target, as when Whitman maps aspects of the spider's thread over to the soul's quest for meaning. Similarly, many extended balance metaphors such as *climate equilibrium* make detailed comparisons, as when breakdowns in climate stability produce tipping points that lead to disastrous floods.

Holyoak notes that analogical resonance in poetic metaphors often transfers emotions. For example, many metaphors about love connect it with emotionally positive concepts such as *rose*, while Shakespeare conveyed caution with his "Love is a smoke made with the fume of sighs." Generally, balance metaphors such as *equilibrium* transfer happiness and goodness about bodies operating in balance, whereas imbalance metaphors involving dizziness, vertigo, and unsteadiness carry with them associated negative emotions such as anxiety.

IMAGE SCHEMAS

Mark Johnson's pioneering book *The Body in the Mind* contains a profound analysis of how bodily experiences of balance extend into metaphors.[2] He emphasizes that balance is an activity that brings experiences such as standing, wobbling, and falling. He proposes that these experiences develop into an image schema, which is a recurring mental structure for perceptual interactions, bodily experiences, and cognitive operations. His basic balance schema consists of a point or axis around which forces or weights are distributed so that they counteract one another. For example, in the balanced budget metaphor, the central point is a zero difference between income and expenditures, which are the weights that need to be distributed. Johnson provides deep discussions of artistic examples of balance that I return to in chapter 9. He identifies three special cases of his balance schema that apply to particular targets: standing or walking without falling, carrying an equal load in each hand, and homeostasis in our body organs, such as managing gas in our stomachs.

I agree with Johnson that the body is the ultimate origin of balance metaphors, but not all balance metaphors use the body as the direct source. Examples such as *balanced budgets* are more similar to the balance scale than to bodily activities such as staying upright. No one knows how the balance scale was invented around 5,000 years ago, but it is plausible that it grew out of the practice of comparing the weights of objects by holding them in opposite hands and feeling the effect on bodily balance. Nevertheless, the balance scale was so socially important that it became as metaphorically impactful as the bodily original. In the age of digital scales, most people are much more familiar with balancing their bodies than with balancing pans.

Johnson's balance schema with a point and counteracting weights covers both bodily balance and weight scale balance, and we will have to investigate whether particular metaphors such as *balanced budget* and *climate equilibrium* are closer to body or scale sources. Moreover, we need a fuller account of how minds produce and imagine activities such as controlling the body, weighing objects, and balancing budgets.

Chapter 1 depicted a metaphor from the Egyptian *Book of the Dead* about weighing the soul against the feather of truth to determine immortality. Such metaphors go far beyond embodiment to deal with nonobservables such as the soul, truth, and immortality. Similarly, Whitman's poem about the spider deals with the transcendent musing soul. Scientific metaphors sometimes also go beyond what is observed, as in discussions of black holes and atomic bonds. Such metaphors are better described as "transbodied" rather than embodied because they transcend the senses. We will see examples of transbodied balance metaphors in science, medicine, society, and philosophy, in applications such as the scales of justice and work-life balance.

At first glance, there seems to be a conflict between Holyoak's account of metaphor as a comparison between a target and a specific source and Johnson's account of metaphor as the application of an image schema. The conflict disappears when we recognize that their accounts reflect different stages in the development of metaphor, which begins with the specificity and novelty that Holyoak identifies in poetry and abstracts into the schemas that Johnson discusses.

SIMULATING EMBODIMENT

Since Johnson's book was published, ideas about the embodiment of cognition have become widely influential in psychology and philosophy. A major factor has been Laurence Barsalou's research on the perceptual basis of linguistic symbols.[3] Concepts like *cat* are not just parts of language but also operate in the mind through simulations of sensory, motor, and emotional states. Your concept of cats includes what they look like, how they sound, how much you like them, and actions you can do with them such as petting. Imagining what to do with them requires simulating these experiences, as in Carl Sandburg's poem that begins with the metaphor "The fog comes on little cat's feet." To map the fog target onto the cat source, we need to imagine the movement of the cat.

Both the body movement and weight scale sources for balance metaphors require simulations that can be multimodal in that they involve different kinds of perception, movement, and emotional feeling. How the brain performs such simulations is best explained by Chris Eliasmith's theory of semantic pointers that I introduced in chapter 4. Simulations can be performed by sequences of if-then rules, as when you imagine how to cook an omelet by applying the rules in box 5.1. These rules can serve as instructions both for actually making the omelet and also for imagining making an omelet by running a mental simulation.

BOX 5.1 Rules for making an omelet

If the eggs are whole, then crack them into a bowl.
If the eggs are in the bowl, then whisk them.
If the pan is cold, then heat it.
If the pan is hot, then melt butter in it.
If the butter is melted, then pour the eggs in.
If the eggs are firm, then add cheese.
If the cheese is melting, then fold the eggs.

As Barsalou emphasizes, such rules are not just verbal representations. The concept *egg* involves sight, touch, and smell. The operation of cracking an egg is a motor action accompanied by sight, touch, and sound. Semantic pointers are neural representations that retain aspects of these sensorimotor inputs, so we can reinterpret rules as having the nonverbal structure <*semantic pointer 1*> → <*semantic pointer 2*>, for example, <*egg*> → <*crack*>. Running a simulation of making an omelet is sequentially applying such multimodal rules. Brain-scanning research on procedural knowledge such as tying different kinds of knots involves frontal, parietal, motor, and cerebellar regions.

Simulated embodiment is important for many balance metaphors, such as the problem faced by government leaders in dealing with COVID-19. People carrying a heavy load in each hand are familiar with the physical process of keeping upright while balancing both loads. Similarly, leaders are faced with the dynamic problem of going forward while balancing lives and livelihoods.

FEATURES OF METAPHORS

Characterizations of metaphorical thinking identify typical features of metaphor that apply to the balance metaphors presented in chapters 6–10. For each metaphor, we can ask about its target, source, comparison, purpose, dimensionality, dynamic properties, conventionality, emotional content, and evaluation. These features also apply to countless domains of metaphor besides balance, and identifying them allows us to systematically compare the different metaphors found in science, medicine, society, the arts, and philosophy. Each of these features generates questions that show how the metaphors work.

Target

What is the metaphor aimed at illuminating? For example, the *balanced budget* metaphor concerns the economic practice of making a personal or organizational budget. Targets of balance metaphors arise in every human enterprise as concepts, activities, puzzling occurrences, or problems to be solved.

Source

What concepts are used in the source? Candidates include concepts that assert balance, including *balanced, equilibrium, equipoise, stable*, and *steady*. Contrasting concepts that assert imbalance include *unstable, disequilibrium, unstable, unsteady, shaky, wobbly, dizzy*, *vertigo*, *fall*, and *tipping point*.

What is the size of the source? In a focal metaphor, the source is a simple concept, as in the characterization of the basketball player Fred VanVleet as "steady Freddy." In elaborate metaphors, the source contains much more information that can be used as an analogy for the target.

What modalities are used in the representations of the source? The modality could be words or a variety of sensory images such as pictures, sounds, touches, smells, tastes, and body location, orientation, and motion. For example, the psychological metaphor of falling in love can invoke sensory images that are visual and kinesthetic. Sensory modalities point to embodied sources, but verbal metaphors such as weighing the soul can be transbodied.

Comparison

What are the correspondences between the source and target? Superficial metaphors merely take properties of the source and apply them to the target, for example, if the metaphor *love is a rose* merely means that both are pleasurable. Richer metaphors map relations in the source to the target, as in Shakespeare: "All the world's a stage, and all the men and women merely players." To understand strong balance metaphors such as *immune system balance*, it helps to make explicit the source and the mappings that connect it to the target.

Identifying correspondences works through simultaneous satisfaction of three constraints: meaning, structure, and purpose.[4] The mapping between source and target works most easily when they overlap in meaning, as when both love and roses develop. Mappings also benefit from structural similarity, as when the relation of a person picking a rose corresponds to the relation of a person hurting a relationship. The final constraint is purpose, in that the correspondence should contribute to the goals of the metaphor, such as providing romantic advice. So the analogical mapping that underlies complex metaphors is another kind of sensemaking by constraint satisfaction.

Dimensionality

How many dimensions are used in the comparison between source and target? The simplest balance metaphors are one-dimensional comparisons to weight scales. Scales compare different objects on the single dimension of weight, with the scale tipping toward the pan with the heaviest contents. The metaphor *balanced budget* is one-dimensional because it merely concerns quantities of money.

In contrast, chapter 2 described how bodily balance is multidimensional, integrating ear-based inferences about movement in three dimensions (up-down, back-forth, left-right) with information about the location of the body and about eye movements. Similarly, some metaphors involve mapping between multidimensional sources and targets, as when climate balance is connected to dimensions such as temperature, greenhouse gases, and storms.

Some multidimensional metaphors compare patterns of conscious experiences in the source and the target. For example, when vertigo is used as a metaphor for romantic infatuation, the comparison concerns the experiences associated with falling in love, such as confusion, with the experiences associated with vertigo. Other metaphors, such as the physical and medical ones described in chapters 6 and 7, make no reference to conscious experience.

Purpose

What is the intended purpose of the metaphor? Like analogies, metaphors can serve a variety of cognitive and social purposes, including explanation, problem solving, persuasion, aesthetics, and entertainment. For example, the extensive use of equilibrium concepts in economics is intended to be explanatory and predictive, but the metaphor *economic vertigo* is mostly amusing. Identifying the purposes of a metaphor is crucial for evaluating it as strong, weak, or failed. The purpose of poetic metaphors such as Whitman's spider can range from aesthetic enjoyment to a philosophical explanation of how life ought to be lived.

Conventionality

Is the metaphor novel or conventional? Dedre Gentner and Brian Bowdle analyze the "career" of metaphor (itself a metaphor).[5] They describe how a metaphor evolves from its original, novel use that generates a new comparison between a previously unconnected source and a target. Repeated use allows the source to become more abstract and schematic so that it functions as a standalone concept. According to the Google Books Ngram Viewer, the terms "balanced budget" and "balanced diet" were introduced around 1900 through novel connections with the concept of balancing. The subsequent wide use of these metaphors has rendered them conventional so that people are unaware of their connection with balancing.

Dynamics

Is the metaphor static or dynamic? Simple focal metaphors such as *balanced photograph* can be relatively static, but most balance metaphors are dynamic, in accord with the ongoing experience of balancing the body by constantly making readjustments through body and eye movements. Simulations using multimodal rules enable minds to keep up with the dynamics of rich metaphors, as when maintaining equilibrium in economic systems requires frequent financial adjustments.

Emotion

What is the emotional content of the metaphor? Dry metaphors such as *balanced equation* are devoid of emotional content, but others aim to transfer positive aspects of balance such as security and confidence, as when a *balanced diet* is recommended. In contrast, other metaphors aim to transfer negative aspects of imbalance such as danger and anxiety, as when a buyer is warned about a *shaky deal*. All metaphors with positive or negative emotional value invoke consciousness.

Emotional transfer is clarified by a tool I invented called cognitive-affective maps, as shown in figure 5.1.[6] These maps use ovals to depict concepts with positive values and hexagons to depict concepts with negative values. Straight lines indicate

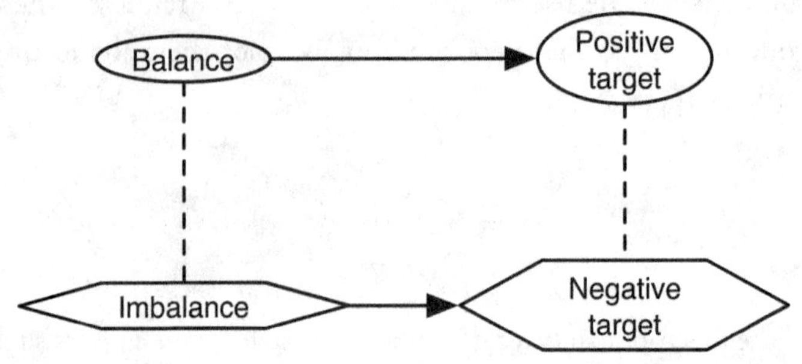

FIGURE 5.1 Transfer of the emotional values from a balance source to a target.

mutual support while dotted lines indicate incompatibility. The arrows indicate transfer of emotional values from the balance source to the various targets described in chapters 6–8.

Evaluation

Is the metaphor strong, weak, bogus, or toxic? Chapter 1 described strong metaphors as successfully accomplishing their goals through legitimate sources that map well onto the target. Weak metaphors are less successful because of problems such as dubious sources and confused correspondences. Bogus metaphors look good on the surface, but scrutiny reveals that they fail to accomplish their purposes because of misleading sources, poor mappings, and lack of relevance to desirable goals. Toxic metaphors are even worse because their use is demonstrably harmful to the people who employ them. We can evaluate balance metaphors by asking about the extent to which they accomplish their purposes and the general effects of using them. In addition to noting strong balance metaphors, I will identify as bogus or toxic some popular metaphors such as *balance of nature* and *yin/yang balance*.

OBJECTIVITY AND EMBODIMENT

In 1980, George Lakoff and Mark Johnson published *Metaphors We Live By*, which has been immensely influential, with more than 70,000 citations according to Google Scholar.[7] They argued that metaphors are more than devices of rhetorical language and contribute pervasively to thought and action. My examination of the influence of balance metaphors supports many of their claims about metaphorical cognition, but it challenges other claims about the embodiment and nonobjectivity of thought.

Lakoff and Johnson propose in their 1980 book, and in their later *Philosophy in the Flesh*, that their views on metaphor imply rejection of dominant philosophical views about objectivity. They summarize: "We have argued that truth is always relative to a conceptual system, that any human conceptual system is mostly metaphorical in nature, and that, therefore, there is no fully objective, unconditional, or absolute truth."[8] They insist that "metaphors are basically devices for understanding and have little to do with objective reality."[9]

In contrast, I argue in my book *Natural Philosophy* that there are good reasons for believing that reality is independent of human thought and that understanding requires grasping reality.[10] Much physical evidence suggests that the universe had existed for more than 13 billion years before minds evolved on Earth. Moreover, no amount of thinking allows ordinary people to see just what they want to see, for example, that their bodies are immune to aging. Similarly, scientists have conceptual systems that guide their experiments, but the results of interactions with the world using instruments such as telescopes often force the rejection of theories and concepts in favor of ones that provide better explanations. Finally, the success of technologies such as electronic devices and disease vaccines shows that science often succeeds in accurately describing the world.

Lakoff and Johnson provide no evidence for their claims that concepts and conceptual systems are mostly metaphorical. Metaphors have undoubtedly been important in the history of thought, as in the examples of natural selection and light waves. However, in my examination of 200 great scientific discoveries and technological inventions, I estimated that only about 15 percent of breakthroughs were based on analogies.[11] Our best current scientific theories include Einstein's relativity, quantum mechanics, atomic and subatomic chemistry, natural selection, and genetics. These powerful conceptual systems have metaphorical aspects, but they also thrive through mathematics and mechanisms that enable them to explain many observable phenomena. An examination of 47,000 words from the British National Corpus found that 13.6 percent of all words related to metaphor.[12] Metaphor is an important contributor to language and thought, but no evidence supports the claim that they are predominantly metaphorical.

Lakoff and Johnson insist on approaching the world from a direction that is "experientalist" and "embodied," but experience and embodiment have limitations. Empiricists such as Locke, Hume, and the logical positivists thought that all knowledge depends on sense experience, but the philosophy of science has established that the most valuable theories go beyond the senses to defend the existence of such nonobservable entities as gravity, atoms, quantum wave-particles, molecular bonds, genes, and the rules and concepts that operate as mental representations. Belief in the reality of these entities is justified by the explanatory power and technological applications of the theories that propose them.[13] For example, we are fully justified in believing in the reality of viruses because they explain diseases that can be prevented by vaccines.

Embodiment is also limited as a source of knowledge. My book *The Cognitive Science of Science* distinguishes between *extreme* embodiment, which understands thinking as just embodied action, and *moderate* embodiment, which allows that language and thought are often shaped but not fully determined by embodied action.[14] The moderate view is well supported by the abundance of evidence that concepts such as *color* and *force* depend heavily on the kinds of bodies we have, but it allows for the operation of transbodied concepts such as nonobservable entities in science, mathematics, and religion.

Lakoff and Johnson make much of the embodied nature of our concept of cause, which is tied in with early experiences of sensorimotor manipulations. But in scientific hands the concept of cause has been enriched to include additional transbodied considerations that include statistical dependencies, causal networks, and underlying mechanisms.[15] My analysis of balance metaphors recognizes their moderate embodiment arising from vestibular and other bodily systems, but also acknowledges that they often abstract far beyond the body in examples such as *economic equilibrium*. More than half of the balance metaphors in this book are not directly connected to bodily balance. Claims that the body shapes the mind underestimate the capacity of neural mechanisms to generate representations and inferences that transcend the senses.

Lakoff and Johnson claim that the idea of objective truth is socially dangerous because it allows people in power to impose their metaphors on the whole culture. On the contrary, truth and objectivity are essential for challenging oppressive views such the racist and sexist metaphors that I branded as toxic in chapter 1. Embodiment and experience are not sufficient to distinguish between medically legitimate balance metaphors such as *immune system balance* and scientifically bankrupt metaphors such as *ying/yang balance*. Sorting out the difference requires the scientific method of evaluating competing theories based on carefully collected evidence, as I describe in chapter 7. Embodiment and experience will not by themselves tell us how to explain and treat diseases effectively.

That science is a better guide to the world than experience is clear from the development of many mistaken views. Everyday perceptual experience suggests that reality accords with Euclidean geometry, but general relativity uses non-Euclidean views about the curvature of space to provide a better explanation of the universe. Some politicians such as Donald Trump think that their feelings should triumph over facts, for which they get to create their own alternatives, but this preference for personal experience over evidence is impotent to deal with medical realities such as pandemics and social realities such as poverty. Some religious people base their faith on their experiences of awe and connection to God, but they cannot explain why their gods are preferable to those of other religions or to none at all. Some pet owners experience their pets as equivalent to people despite the substantial differences that I describe in my book *Bots and Beasts*.[16]

Understanding can be illusory if it is based on experiences, metaphors, and concepts that do not stand up to scientific scrutiny. Similarly, claims that the mind extends into the world are helpful when they recognize that thinking benefits from interactions with bodies, physical space, technology, and other people. But all of these extensions depend on neural mechanisms for perception, problem solving, and learning that are the core of mind.

Because of the limitations of experience and embodiment, I will use scientific methods of evidence and theory to evaluate balance metaphors. Transcending experience and embodiment provides both a better grasp of reality and a greater ability to change it in line with real human needs. Abandoning objective truth favors autocrats, quacks, and con artists instead of the vast majority of people. Experience and embodiment cannot tell you what is wrong with vaccine hesitancy, political conspiracy theories, and climate change denial.

COMBINING METAPHORS AND MECHANISMS

The nine features of metaphor (target, source, comparison, dimensionality, purpose, conventionality, dynamics, emotion, and evaluation) provide tools for analyzing and comparing a host of balance metaphors in many domains. They also provide a deeper characterization of the concept of metaphor than its simple definition as a linguistic comparison between a source and a target, which by itself does not distinguish metaphor from simile and analogy. Chapter 4 analyzed consciousness using exemplars, typical features, and explanations, and the same method applies to metaphor. We have already seen good examples of metaphor, with many balance examples to come. The nine features of metaphor I presented provide a good start on the typical features of metaphor, which are not intended to be necessary and sufficient conditions in the spirit of classical definitions. For example, a linguistic expression can be a metaphor even if it lacks emotional content or evaluation, but these features apply to many metaphors in addition to the more standard features.

What does the concept of metaphor help to explain? Without it, we would be puzzled that so much of language makes statements that are not literally true. Identifying metaphors is also important for explaining how nonliteral language can sometimes be more effective than more straightforward language for purposes such as literary expression and persuasion. In turn, we want to be able to explain how metaphor works using mental mechanisms that accomplish mappings between source and target.

Routine, focal metaphors (sometimes called "dead" metaphors, which is itself a metaphor) may not add much to the understanding of a target domain, but some rich metaphors add enormous analogical resonance. Keith Holyoak points out the irony of Amos Tversky's dismissal of the use of metaphor as a "cover-up," which is itself metaphorical. Winston Churchill's remark that democracy is the worst form of government except for all the others can be adapted to say that metaphor is the worst way of thinking except for all the others.

This chapter serves as a bridge between the neural accounts of balance, vertigo, and consciousness in chapters 2–4 and the extensive discussions of balance metaphors in chapters 6–10. The neuroscience chapters expand on the common experiences of balance to provide rich source analogs for important targets in science, medicine, society, the arts, and philosophy. Figure 5.2 sketches the genealogy of balance metaphors, showing the conjectured historical relations between some of them. The connection between literal balance and the many balance metaphors to come is more than use of the word "balance," thanks to the transfer of aspects of human balancing to weight scales and many other practices.

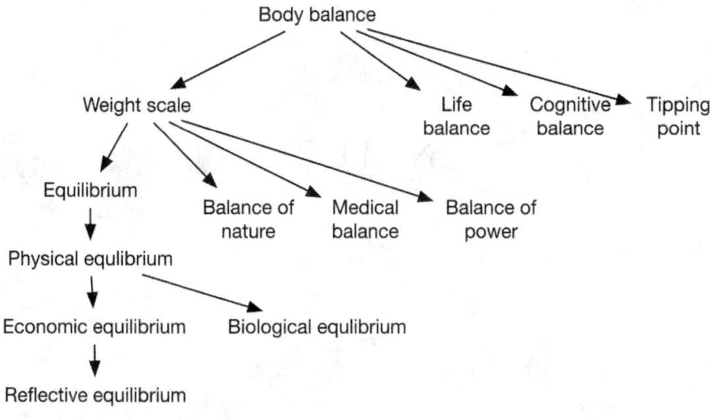

FIGURE 5.2 Genealogy of balance metaphors. Arrows indicate the likely analogical sources for the targets.

I have not abandoned mechanisms for metaphors, because I rely on neural and cognitive mechanisms for explaining how metaphor works. The ultimate goal of this investigation is to combine mechanisms and metaphors to show how balance contributes to human thinking through varieties of sensemaking.

6

NATURE

Balance metaphors abound in the natural and social sciences, especially through the concept of equilibrium, which is important in physics, chemistry, biology, economics, and sociology. Some of these metaphors, such as *balanced equations*, are one-dimensional and based on weight scales, but the biological and social sciences also employ multidimensional balance metaphors more analogous to the brain's balance system. For example, the *balance of nature* concerns the distribution of species in an ecosystem, where predators and prey constrain each other's existence. Because wolves kill and eat elk, more wolves in an environment make for fewer elk, but fewer elk lead to fewer wolves and then more elk, assuming that neither wolves nor elk have become extinct. Ecological equilibrium occurs when the numbers of wolves and elk are balanced against each other.

We can ask whether balance and equilibrium are normal in two senses. The term "normal" can just mean "usual," which is the descriptive assessment of whether balance is statistically common. Or it can mean "desirable," which is the prescriptive judgment that balance satisfies appropriate goals. Bodily balance is normal in both senses: usually people manage without falling down, and vertigo is uncommon; stability is desirable because it enables people to accomplish goals such as getting around in the world.

The evaluation question is whether metaphors such as *balance of nature* and *chemical equilibrium* contribute to scientific explanations or impede them. In ecology, the balance metaphor has been criticized and has been supplanted by a more value-neutral equilibrium metaphor, which is also challenged by imbalance metaphors such as *tipping point*. Before getting to the use of balance ideas in biology, let us consider their use in mathematics, physics, and chemistry. In all these fields, balance metaphors contribute to sensemaking.

MATHEMATICS

Mathematics is sometimes construed as an abstract enterprise independent of nature, but I prefer the view that math is the science of quantities and structures in the world.[1] Mathematics is usually assumed to be rigorous and literal, but metaphor and analogy make large contributions to mathematical understanding and creativity. George Lakoff and Rafael Núñez documented many cases where mathematics develops through mappings between familiar source domains such as geometrical space and abstract structures.[2] For examples, we can understand integers and real numbers as points on a straight line.

Algebra uses an important metaphor for the method of balancing an equation that was invented in Baghdad by Muḥammad ibn Mūsā al-Khwārizmī. Around 820 AD, he wrote the first algebra text, *The Compendious Book on Calculation by Completion and Balancing.*

If you are asked to solve the equation $4x - 3 = 2x + 5$, you balance it by subtracting $2x$ from each side and adding 3 to each side, yielding $2x = 8$. This operation is analogous to taking equal weight off both sides of a two-pan weight scale. The next operation is to divide both sides by 2, which yields the answer $x = 4$, where dividing by 2 amounts to subtracting half and therefore also corresponds to removing weight from a scale. For scales, the only dimension that matters is weight, and for balancing equations the only dimension that matters is numerical quantity. So balancing an equation is a one-dimensional metaphor.

Is the scale source embodied or transbodied? The weight scale is an object that can be physically manipulated by pushing the pans up and down, so embodiment contributes to the appreciation of the source, especially if the weight scale is associated with the bodily act of comparing the weight of different objects by holding them in different hands. But the numerical aspects of the weight scale introduce the transbodied aspects of numbers in that they transcend mere observations of things, especially for large numbers. Moreover, the abstractness of the variables x and y in equations requires transbodied representations. So the metaphor *balanced equation* is both embodied and transbodied.

The scale source and the equation target are dynamic in that it can take a series of operations of adding, subtracting, multiplying, and dividing to get the desired result. Weight scales are also sometimes used to help children understand how fractions work, for example, that $\frac{1}{4} + \frac{1}{4} = \frac{1}{2}$.

The different operations and multiple steps show that *balanced equation* is not just a focal metaphor that can be understood by combining the concepts *balance* and *equation*. Rather, it is an extended metaphor that works through the analogical mapping shown in table 6.1. The expression "balancing an equation" is in such routine use that people may not be aware of the analogy that inspired it and makes it work. Many dead metaphors like

TABLE 6.1 Analogy underlying the metaphor of balancing equations

Weight scale (source)	Equation (target)
Two pans	Two sides of equation
Weight on each pan	Numerical quantity on each side
Balanced when equal	Balanced when equal
Adding weights	Adding quantities
Removing weights	Subtracting quantities
Adding weights repeatedly	Multiplying quantities
Removing weights repeatedly	Dividing quantities
Adding weights to both pans retains balance	Adding, subtracting, multiplying, or dividing both sides keeps them equal

kick the bucket have become so familiar that they have lost their original imagery. Balancing an equation is almost as familiar as kicking the bucket but is still enriched by appreciating the original mapping used when the metaphor was novel.

Even mathphobics do not succumb to dizziness or vertigo when an equation needs balancing, so no strong emotion is associated with the mathematical balance metaphors. On the other hand, since solving a mathematical problem using equations can be pleasurable, for some people the metaphor of *balancing equations* is moderately positive.

How good is the metaphor of *balancing equations* for accomplishing its intended purpose of enabling people to solve algebraic problems? The web contains educational sites that use pictures of weight scales to explain how to balance equations, so the metaphor continues to be helpful and deserves to be evaluated as strong.

PHYSICS

The first balance scales were invented for practical purposes more than 4,000 years ago in Egypt and in the Indus Valley of what is now Pakistan. The ancient Greeks later invented balances with unequal arms, which contributed to early physics by inspiring the law of the lever.[3] A balance scale is in equilibrium when the weights on its two pans and the length of its arms result in there being no change in the movement up and down of the pans. This concept of equilibrium as a stable state resulting from balanced forces has become important in diverse fields.

In physics, forces are balanced when they are of equal magnitude and opposite directions, as when the force of gravity pulling your body down is compensated for by the force of the ground pushing it up. Equilibrium occurs when the compensating forces ensure that no acceleration takes place. The metaphor of *balanced forces* is multidimensional because it concerns not just the amount of force but also one or more directions of force that can cause complications, for example, if three people are all pulling on a blanket. In material objects, the positive protons and negative electrons attract and repel each other, but the forces are balanced so we do not feel them.

In the nineteenth century, balance became an even more important concept in physics because of theories about heat equilibrium. Two systems are in thermal equilibrium when they are connected but have no flow of heat between them. Just as a weight scale is in equilibrium when the two pans have the same weight, so rooms are in thermal equilibrium when the two rooms have the same temperature. This one-dimensional metaphor is dynamic because temperatures can change as heat flows back and forth between rooms to restore equilibrium. Table 6.2 lays out the correspondences that concern relations as well as properties.

TABLE 6.2 Analogy underlying the metaphor of thermal equilibrium

Weight scale (source)	**Heat (target)**
Two pans	Two areas
Weight on each pan	Temperature in each area
Balanced when weights are equal	Equilibrium when temperatures are the same
No movement of balanced pans	No temperature change in equilibrium

Thermal equilibrium changed from mere metaphor to mechanism when Ludwig Boltzmann and others came up with mathematical descriptions of heat exchange based on the motions of atoms and molecules. Before the development of this kinetic theory of heat, heat was understood as a fluid substance called "caloric" that moved between locations until equilibrium was achieved. Neither the element caloric nor molecules in motion are observable by the senses, so the *thermal equilibrium* metaphor has a transbodied character that goes beyond the embodiment of the pan balance.

Emotion and consciousness are not major players in thinking about thermal equilibrium, although finding an explanation of the behavior of heat brings some people a modicum of pleasure. The metaphor of thermal equilibrium serves its purposes well, both as providing explanations and as suggesting a mathematical tool for solving practical problems. One hundred and fifty years of usefulness mark the metaphor as strong.

Thermal equilibrium only concerns heat energy, but the later idea of thermodynamic equilibrium is a broader concept that also considers chemical and mechanical energy, making it a multidimensional metaphor. The laws of thermodynamics are fundamental principles of physics that characterize how systems change in and out of equilibrium. The first law of thermodynamics states that total energy can neither be created nor destroyed, so that different forms of energy and matter balance each other out.

Thermodynamic equilibrium is too complex for the weight scale to serve as the source and is more analogous to the brain's balance system, which integrates multiple dimensions. However, the originators of thermodynamics probably got their ideas by generalizing from thermal equilibrium rather than by using an analogy with bodily balance. Therefore, the metaphor of *thermodynamic equilibrium* most likely had its origin in the already abstract, transbodied, and dynamic source metaphor of *thermal equilibrium*.

With the concept of equilibrium, we see a metaphorical chain reaction from bodily balance to scale equilibrium to other kinds of equilibrium in physics, chemistry, biology, economics, and philosophy.[4] Here the concept of chain reaction serves as a metaphor by analogy to nuclear physics. Another metaphorical chain reaction in the history of science concerns waves. The ancient Greeks developed the wave theory of sound by analogy to water waves, and seventeenth-century Europeans formed the wave theory of light partly by analogy to sound waves.[5]

CHEMISTRY

Chemistry uses balance metaphors for both its equations and its mechanisms. A chemical equation is different from an algebra equation because it describes a chemical reaction rather than a numerical equality. For example, the reaction that combines aluminum and oxygen to produce aluminum oxide is:

$4Al + 3O_2 \rightarrow 2Al_2O_3$. Balancing requires having the same number of atoms of each element on both the left and right sides of the formula. Balancing a chemical equation is analogous to balancing an algebraic one because multiplication can be used to make the correspondences work. As with algebra, only one dimension matters—the number of atoms. The nineteenth-century metaphor of balancing chemical equations did not depend on the balance scale as its source because it could rely on the familiar metaphor of balancing equations in algebra, another metaphorical chain reaction.

Whereas balancing chemical equations is a one-dimensional metaphor, the metaphor *chemical equilibrium* is like *thermodynamic equilibrium* in being mechanistic and multidimensional.[6] Chemical equilibrium requires a balance among different substances in chemical reactions, which are processes that transform reactants (substances consumed) into resulting products. For example, sodium and chlorine combine to produce sodium chloride. Chemical reactions are mechanisms in which the parts are chemicals that are changed by interactions among their atoms and electrons. A reaction is in chemical equilibrium if the concentrations of reactants and products have no further tendency to change. For example, acids and bases can be stable independently but violently lose equilibrium when combined, for example, if you mix vinegar and bleach. Because reactions involve at least two reactants and products, the metaphor of *chemical equilibrium* is multidimensional.

The idea of chemical equilibrium arose around 1790 in the work of a French chemist, Claude Louis Berthollet. He realized that the received view that chemical combinations result from the affinities between substances could not account for quantitative differences in chemical reactions. Instead, he drew from Newtonian physics the idea that reactions result from the

TABLE 6.3 Physics analogy underlying the metaphor of chemical equilibrium

Mechanical equilibrium (source)	**Chemical equilibrium (target)**
Objects exerting force on each other	Reactants and products interacting
Unbalanced forces cause motion	Unbalanced reactants cause reactants to produce products
Balanced forces produce no change	Balanced reactants produce no change

interactions of forces. As with mechanical equilibrium, chemical equilibrium results when the balancing of forces stops change.

Thus the metaphor of *chemical equilibrium* is based on an analogy with physics rather than with bodily balance or the balance scale. Table 6.3 shows the mapping, with chemical reactants corresponding to objects whose forces produce change. In both cases, balanced forces lead to equilibrium.

The chemistry metaphors of *balancing equations* and *chemical equilibrium* have little impact on emotions and consciousness. But they still contribute to problem solving and explanation in scientific thinking and education and accordingly deserve to be evaluated as strong.

BIOLOGY

In the Disney movie *The Lion King*, the father lion tells his son that everything he sees exists in a delicate balance. This metaphor of the balance of nature has ancient origins and persists in current environmentalist concerns such as David Suzuki's book

The Sacred Balance. Contemporary ecologists are skeptical of the value of this way of thinking, but they retain another balance metaphor that takes equilibrium among organisms to be a natural state of ecosystems. More radically, some biologists advocate a shift toward imbalance metaphors such as *tipping point*. Other balance metaphors in biology include the application of equilibrium to genetic variation and the view that organisms operate with mechanisms of homeostasis that return them to stable states. We need to understand the mental structure of such metaphors and evaluate whether they aid or hinder the understanding of nature.

The *Lion King* idea that nature is balanced has ancient origins, as much of early Greek science assumed that nature is constant and harmonious. Credit for the balance among different kinds of animals was due to the gods, who could be encouraged by prayer, rituals, and sacrifices. Maintaining balance could avoid undesirable forms of change such as the extinction of kinds of animals. Greek philosophy before Socrates had an essential tension between advocates of constancy (Parmenides and Pythagoras) and advocates of change (Heraclitus and Empedocles) who maintained that everything is in a state of flux.

The ancient ideas about the constancy of nature are less a scientific theory than a coherent set of values that can be displayed using a cognitive-affective map. Figure 6.1 shows part of the value system of the ancient Greeks, with positive values such as *balance* indicated by ovals and negative values such as *change* indicated by hexagons. Mutually supported values are connected by straight lines, and incompatible values are shown by dashed lines. The map in figure 6.1 shows how the balance of nature ties together a host of values of the ancient Greek worldview, including the contribution of the balance of humors to health (chapter 7) and the importance of harmony in music (chapter 9).

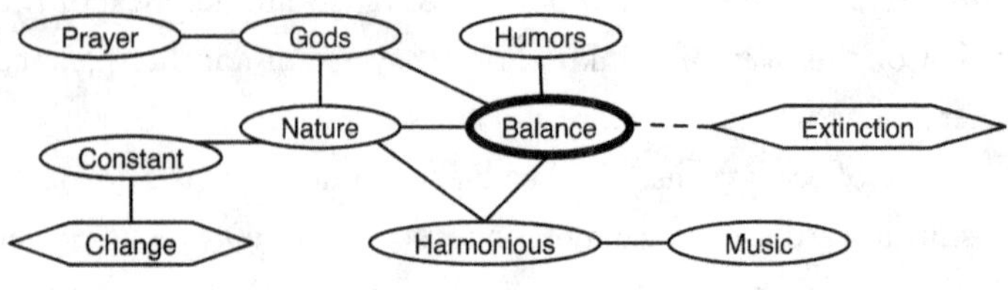

FIGURE 6.1 Cognitive-affective map of ancient Greek values about the balance of nature.

In the eighteenth century, the idea of the balance of nature became less theological and involved detailed studies of interacting species such as insects and agricultural plants.[7] The Swedish professor of natural history, Carl Linnaeus, began the serious study of ecology by writing about the "economy of nature" that was maintained by the propagation, preservation, and destruction of species. Nineteenth-century biologists increasingly recognized the role of competition in maintaining proportional balance among species, culminating in Charles Darwin's theory of evolution by natural selection. *Selection* is a metaphor based on the analogy between competition among species and the selection of breeds carried out by human breeders. Darwin discussed examples of balanced numbers such as between bees and clover and between cats and mice. The assumption remained that natural selection and other processes supported the basic stability of nature.

What kind of balance is the source for the metaphor *balance of nature*? Table 6.4 takes a first pass at a mapping in a simple case of ecological ıbalance involving just two species, wolves and elk. At first glance, balancing the number of wolves and elk is like balancing the weights on the different pans in a weight scale.

TABLE 6.4 Analogical basis for the metaphor *balance of nature*

Weight scale (source)	**Nature (target)**
Two pans	Numbers of wolves and elk
Weight on each pan	Number of wolves on one side and number of elk on the other
Balanced when equal	Balanced when appropriate number of each
Adding weights to one side	Adding wolves or elk
Removing weights from one side	Subtracting wolves or elk
Adding weights to one pan makes it go down and makes the other pan go up	Increasing the number of wolves makes the number of elk go down, and decreasing the number of wolves makes the number of elk go up

With closer scrutiny, this analogical mapping has several problems. Unlike the weights on the two pans, the number of wolves and elk does not have to be exactly equal, just appropriate for the environment. Moreover, increasing the number of elk would not decrease the number of wolves, who would then have more to eat. Most importantly, wolves and elk are not the only organisms in the ecology and their numbers depend on other factors, such as the availability of food sources like rabbits for wolves and plants for elk.

Perhaps we could get a more coherent mapping by comparing the balance of wolves, elk, and other species with the multidimensional problem of balancing the body by reconciling information from diverse senses and the brain's expectations. But that comparison affords no clear comparisons between numbers of species and sensory inputs. At best, the *balance of nature*

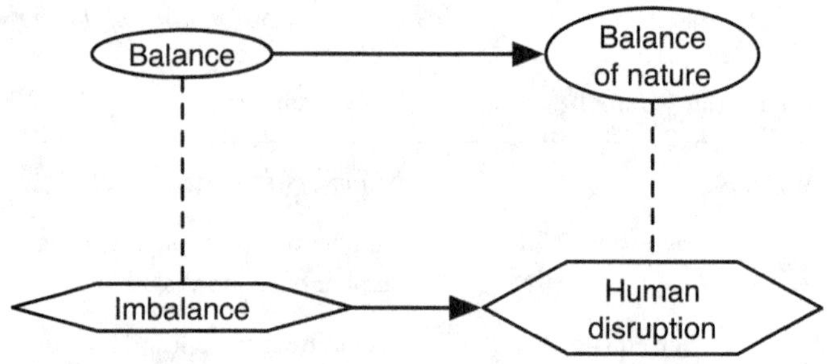

FIGURE 6.2 Transfer of emotion from bodily balance to balance of nature. Ovals indicate positive values, while hexagons indicate negative values. Dotted lines indicate incompatibility, and lines with arrows indicate emotional transfer from source to target.

metaphor merely suggests an emotional connection between the positive value of bodily stability and some corresponding ecological value. Figure 6.2 shows the emotional analogical transfer from body balance to balance of nature, including the negative quality of human disruption.

Contemporary environmentalists and conservationists maintain that the balance of nature is desirable. The threat comes from human activities that are increasingly driving other species to extinction. Humans disturb the balance of nature through habitat disruption, direct destruction by fishing and hunting, chemical pollution, introduction of invasive species, and global warming.[8] The Earth has had five mass extinctions that each eliminated at least 75 percent of all species, and evidence is mounting that human activity is causing a sixth mass extinction. This configuration of values is shown in figure 6.3. Positive values such as balanced nature are threatened by negative human activity leading to mass extinctions.

Concerns about mass extinctions are scientifically legitimate, but in the twentieth century the concept of balance of nature

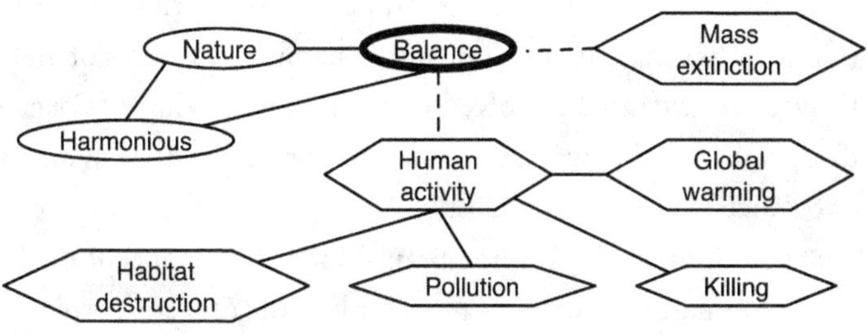

FIGURE 6.3 The balance of nature as threatened by human activity. Ovals indicate positive values, while hexagons indicate negative values. Solid lines indicate mutual support, while dotted lines indicate incompatibility.

became increasingly suspect. Reasons for doubting whether the balance of nature is a reasonable part of ecology include the following:

- Extinction of species is much more the norm than the exception: around 99 percent of all the 5 billion species that have ever existed have gone extinct, and at least five mass extinctions have occurred. The typical duration of a species is only about a million years.
- Even without extinctions, the numbers of animals are constantly varying with dramatic irregularities. Ecologies may have stability over short time spans, but over large periods they display much volatility.
- No mechanisms (scientific or theological) support balance, and natural selection can just as easily lead to extinctions as to stability.

Hence balance does not seem to be a usual part of nature, and no good reason asserts its inherent desirability.

If the dinosaurs had not become extinct during the fifth (Cretaceous) mass extinction, small mammals would not have thrived and eventually evolved into the most intelligent species the planet has known. Human evolution depended on many unbalanced events, including the development of species of primates with increasingly large brains. Hence the concept of the balance of nature has lost its place both at the center of explanations of nature and at the center of values about nature. Nature is not inherently a beneficent force. *The Lion King*, the ancient Greeks, early biologists, and some recent environmentalists are wrong. John Kricher remarks: "Always more of an assumption than a demonstrated scientific principle, the balance of nature remains in the minds of many people an uncritical paradigm for how they view nature."[9]

The *balance of nature* metaphor fails to accomplish both of its main purposes, explanation and persuasion. It fails at biological explanation because it provides no mechanism to account for the variations in species numbers in an ecology, and it fails at persuasion because it provides no good reasons for caring about the stability and survival of species of organisms. Alternatively, supportable reasons for responsible maintenance of ecosystems include meeting human needs for food and climate stability, scientific interest in diverse organisms, and the concern and delight that many people feel for animals and plants. Accordingly, the *balance of nature* metaphor is weak at best and more appropriately deserves to be recognized as bogus. The metaphor does not warrant castigation as toxic because it has not, as far as I know, been used by ecoterrorists to justify harm to people.

In contemporary ecology, the *balance of nature* metaphor has largely been supplanted by the concept of equilibrium, which, like the equilibrium concepts in physics and chemistry, is less value laden.[10] Mathematical ecology develops models in which equilibrium is the absence of change in the population densities of the relevant species. Ecological equilibrium is analogous to chemical equilibrium, with stability in species populations mapping to stability in numbers of chemical reactants.

Such ecological equilibria are not assumed to be either usual or desirable. Computer simulations have shown that plausible assumptions allow large fluctuations in numbers of animals. The concept of equilibrium is not as emotionally charged as the concept of balance, which inherits a positive value from its association with physiological balance and social phenomena such as balanced budgets.

Ecological equilibrium provides the mathematical basis for the concept of *resilience*, the rate at which populations recover their former densities. As a purely mathematical notion, resilience is value neutral, but some environmentalists give it the same positivity as balance of nature. If polar bears are approaching extinction because melting sea ice is making it hard for them to catch seals, people hope that their degraded environment will improve so that their numbers will return to what they once were. Similarly, psychological resilience is a positively valued condition because we hope that people can recover from serious blows such as illness and bereavement. Both ecological and psychological resilience are metaphorically kinds of regaining balance after instability.

Ecological equilibrium is a more exact idea than the balance of nature, but it is still open to scientific challenge on both empirical and theoretical grounds. Stephen Jay Gould argued that we should think of biological evolution as consisting of punctuated equilibria in which long periods of stability are interrupted by dramatic changes such as mass extinctions.[11] More generally, some current biologists emphasize disequilibrium events known as critical transitions. The usual imbalance metaphor applied to such dramatic changes is *tipping point*, which I analyze later in connection with climate.

Besides ecology, concepts of balance and equilibrium have influenced other areas of biology. In genetics, the Hardy-Weinburg equilibrium is an important principle concerning the frequency of alleles (gene variants) in a population of organisms. Before the mathematician G. H. Hardy became interested in the problem, it was widely believed that a dominant allele would increase in number. He showed instead that allele frequencies remain constant across generations without other evolutionary influences such as mutation and natural selection. This result is idealized because in real environments such influences do operate, and Hardy's proof makes unrealistic assumptions, for example, that the population has an infinite size. Nevertheless, the application of the equilibrium metaphor is useful in describing an ideal state in which the frequencies of gene variants are unchanging. The Hardy-Weinberg equilibrium is analogous to ecological equilibrium, with allele frequencies corresponding to numbers of species members. The metaphorical chain reaction traverses these kinds of equilibria: physiological, weight scale, thermal, chemical, ecological, and genetic.

Another equilibrium metaphor is important in physiology for describing how bodies regulate themselves. The term "homeostasis" was coined in 1930 from Greek words for "same" and "state." Organisms can maintain equilibrium in their internal states through feedback processes such as sweating to lower high temperatures and shivering to raise low temperatures. Other homeostatic processes help to regulate blood glucose, concentrations of sodium and potassium, and balance between neurotransmitters such as dopamine and serotonin. Failures of homeostasis can lead to medical disorders such as hypothermia, diabetes, and schizophrenia. Chapter 7 describes how diseases can arise from failures to maintain homeostasis.

So *homeostasis* is a legitimate biological balance metaphor along with *ecological equilibrium* and the *Hardy-Weinberg equilibrium*. These three metaphors contrast with the bogus *balance of nature* metaphor, which provides misleading explanations of species development and inadequate prescriptions for maintaining ecosystems.

CLIMATOLOGY AND TIPPING POINTS

Climatology (also known as climate science) is the scientific study of how weather conditions vary over time. *Climate equilibrium* occurs when patterns of temperature, atmospheric conditions, and weather extremes remain roughly constant. This balance metaphor is multidimensional because it concerns all of these interacting variables. The concept of climate equilibrium became popular in the 1980s, perhaps by analogy to chemical or ecological equilibria.

In recent decades, scientific research has shifted from talking about equilibria to talking about the dramatic climate shifts that are taking place as the result of global warming. These shifts are often described using the imbalance metaphor *tipping point*, which resonates with people because of its association with the experience of tipping over from dizziness or being toppled. Tipping points also occur in physical and chemical systems when equilibrium is replaced by sudden changes such as water freezing or a chemical mixture exploding.

In contemporary discussions of complex systems, *tipping point* is one of many ideas that signal a shift from changes in degree to changes in kind, including *phase change*, *phase transition*, *critical transition*, *nonlinearity*, *inflection point*, *bifurcation*, *catastrophe*, *singularity*, and *discontinuity*. Specific fields use ideas that also indicate changes in kind and not just degree, for example, *punctuated equilibrium*, *ecological threshold*, *regime shift*, *crisis*, *revolution*, *Gestalt switch*, *liminality*, and *paradigm shift*. None of these are balance metaphors, but since 2000 the vivid metaphor of tipping point has become increasingly common, perhaps because of Malcolm Gladwell's popular book *The Tipping Point*.[12]

A careful characterization of tipping points is provided by Manjana Milkoreit and her colleagues: "a tipping point is a threshold at which small quantitative changes in the system trigger a non-linear change process that is driven by system-internal feedback mechanisms and inevitably leads to a qualitatively different state of the system, which is often irreversible."[13] This definition mentions feedback mechanisms without saying what mechanisms are or how feedback is a special kind of interaction. But we can reformulate this characterization in terms of mechanisms and emergence.

Tipping points occur in mechanisms that consist of combinations of connected parts whose interactions produce regular changes, including emergent properties of the whole combination. These properties are not just aggregates of the changes in the parts because they result from interactions among the parts. The parts interact with amplifying feedback loops in which changes in a part cause changes in another part that cause changes in the first part. As a result of these interactions, the whole combination acquires novel, nonaggregative properties whose emergence marks changes as qualitative (matters of kind) rather than quantitative (matters of degree). Tipping points are states in which emergent properties can suddenly occur, just as when a previously stable person falls over because of dizziness, a strong wind, or a forceful punch.

TABLE 6.5 Body analogy underlying the metaphor of tipping points

Body (source)	Tipping point (target)
Body is stable	System is stable
Body is affected by gradually increasing forces	System parts slowly change in magnitude
Forces on body pass a threshold	System changes pass a threshold
Body suddenly falls over	System suddenly changes to another state

Table 6.5 shows the mapping between bodily instability and more general tipping points. The tipping point idea has no explanatory superiority over previously used ideas such as phase transition, but it is more vivid by virtue of embodiment. Some tipping points, such as global warming, are expected to produce changes that are irreversible, whereas bodies that fall down can usually get up. While the *tipping point* metaphor is most generally embodied, particular versions of it employ transbodied concepts, such as the greenhouse gases invoked in the explanation of climate tipping points.

Tipping point metaphors are usually stated in words but sometimes can be more strikingly presented by diagrams that show the relevant interacting parts or by graphs that show sharp changes in quantities such as temperature. These metaphors concern relations between the parts in a system such as the causal interactions that lead to sharp changes. Tipping points are usually multidimensional in that they involve interactions among different factors, for example, atmospheric temperature, ocean currents, and solar radiation, which all affect climate change. Tipping point metaphors are dynamic because they deal with system changes. The idea of a tipping point was novel when it first was used for climate change and other phenomena in the late twentieth century, but it is becoming increasingly conventional from widespread use.

Timothy Lenton and his colleagues argue that the climate system is approaching global tipping points likely to bring irreversible changes, including ice collapse, biosphere boundary disruption such as Amazon deforestation, and global cascades resulting from interactions of ocean and atmospheric circulation.[14] The basic mechanisms of global warming are well understood: "Earth transforms sunlight's visible light energy into infrared light energy, which leaves Earth slowly because it is absorbed by greenhouse gases. When people produce greenhouse gases, energy leaves Earth even more slowly—raising Earth's temperature."[15] Michael Ranney and Dav Clark found that explaining to people this global warming mechanism increases acceptance of claims about human-caused climate change.[16] In this mechanism, the parts include greenhouse gases such as CO_2, the atmosphere, solar radiation, clouds, and the Earth's surface, including seas, glaciers, and other components.

The interactions among the parts are concisely shown by the arrows in figure 6.4, with solar radiation warming the Earth's surface and infrared radiation emitted from the surface but absorbed and reemitted by greenhouse gases. These interactions result in the currently observed changes in global temperatures, which are approaching tipping points that will bring changes in kind that can be understood as emergent properties of the whole system. The most remarkable emergent property of the system is irreversibility, that the system cannot revert to earlier cooler states with fewer extreme events such as flooding. Once submerged, Florida is unlikely to rise again. Hence tipping points dramatically overturn climate equilibrium.

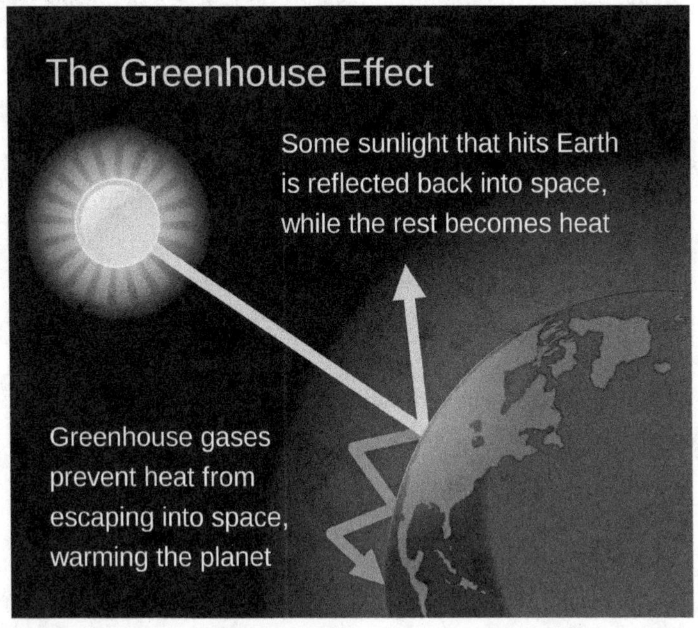

The Greenhouse Effect
Some sunlight that hits Earth
is reflected back into space,
while the rest becomes heat
Greenhouse gases
prevent heat from
escaping into space,
warming the planet

FIGURE 6.4 Simplified mechanism of the greenhouse effect.

Source: Efbrazil/Wikimedia Commons.

Tipping points in climate change are scary because irreversible global warming will afflict many parts of the world with problems such as storms, floods, droughts, and fires. However, tipping points are sometimes good, as with progressive political changes such as the American Revolution and with medical treatments that work suddenly to cure a disease. In 2021, vaccination in some countries produced tipping points in the spread of COVID-19 with dramatic drops in infections. Imbalance metaphors are usually emotionally negative because of the harmfulness of dizziness, vertigo, and falls. But the *tipping point* metaphor is unusual in that it can also be applied to dramatically good changes. Al Gore titled his 1992 book *Earth in the Balance*, where the "in the balance" phrase suggests that there is still time to avoid climate change disaster.

Tipping point metaphors explain past changes such as revolutions and predict future abrupt changes such as climate collapse. They provide a valuable counterbalance to equilibrium metaphors that assume that stability is the natural state of complex natural and social systems. As chapter 10 argues in more detail, thinking about complex systems requires a balance between balance and imbalance. Tipping point metaphors can be misleading for systems that really are stable, but they are effective in characterizing the trajectories of complex dynamic systems such as climate change. Accordingly, they belong in the class of strong balance metaphors in science.

STABILITY AND INSTABILITY

Ancient Greek philosophers set up a rigid dichotomy in understanding the world, between Heraclitus's claim that everything is in flux and Parmenides's insistence that change is an illusion. Modern science recognizes both stability and instability in natural systems, which are conveniently captured by an array of balance and imbalance metaphors. The stable states of physical and chemical systems can aptly be described as being in equilibrium when change does not occur because opposing forces are canceled out. On the other hand, when the balance of forces fails, systems undergo tipping points that produce dramatic changes. Hence both balance and imbalance metaphors contribute to sensemaking in the natural sciences.

This chapter has identified some strong metaphors about nature used in mathematics, physics, chemistry, and biology.

Balance and equilibrium help to explain important natural phenomena, from thermodynamics to homeostasis in organisms. But the limits of balance metaphors are clear in the misleading ecological use of the *balance of nature* metaphor, which is more romantic than scientific. The alternative, imbalance metaphor of *tipping points* provides a more dynamic way of thinking about complex systems that are subject to dramatic changes ranging from climate change catastrophes to ecological collapses.

Knowledge and experience of our bodies contribute to the power of balance and imbalance to illuminate natural phenomena, but metaphor can also transcend the body to introduce nonsensory factors such as chemical reactions and greenhouse gases. Balance metaphors are both embodied and transbodied. Biological balance metaphors about equilibrium and homeostasis carry over into medicine in valuable ways, but the next chapter shows that some medical balance ideas are bogus.

7

MEDICINE

People suffer from diseases caused by balance problems in the inner ear and in the brain, as chapter 3 described when discussing dizziness and vertigo. But long ago the medical traditions in three major civilizations held that *all* diseases result from imbalances in the body's vital substances. Ancient Greek, Chinese, and Indian medicine each claimed that the difference between health and disease is a matter of balance in a small number of factors. Today, we have good reasons to believe that these balance theories are false because scientific theories provide better explanations of the causes and progression of diseases. Nevertheless, the Chinese and Indian traditions survive in their original countries and in the West through alternative medicine.

Balance metaphors have not been banished from scientific medicine. Physicians and researchers legitimately talk about balanced diets, electrolytes, immune systems, and neurotransmitters. Unlike concerns with dizziness and vertigo, where losing balance is literally a problem, these uses are metaphorical but legitimate because they point to evidence-based mechanisms. In contrast, the ancient balance traditions from Greece, China, and India have three major flaws: they are unsupported by carefully collected evidence, incompatible with well-established scientific theories, and lacking in mechanistic explanations of why they are supposed to work.

Nevertheless, a few effective practices in the Chinese and Indian traditions are worth pursuing even though the traditional explanations of their efficacy are bogus. Tai chi and yoga have both proved useful for dealing with balance and other medical problems, but their usefulness can be explained by modern medicine and cognitive neuroscience. Medical sensemaking is essential for human health but needs to be vigilant in its use of balance metaphors.

ANCIENT BALANCE THEORIES OF DISEASE

The oldest medical explanations are theological or magical: diseases are the result of the actions of gods or sorcerers.[1] For example, the ancient Babylonians and Egyptians thought that diseases were signs that the gods were offended and messing with people's bodies. Treatment consisted of rituals of prayer and exorcism to get rid of evil gods or demons. Some indigenous people of North America attributed disease to intrusions in the body by objects placed by sorcerers and sought the help of shamans to remove the objects.

In contrast, nontheological theories of health developed more than 2,000 years ago in Greece, China, and India. All three were balance theories, although the factors they identified as responsible for disease were different. The Greeks pointed to four humors (blood, phlegm, black bile, yellow bile), while the Indian Ayurvedic tradition cited three doshas (vata, pitta, and kapha). The ancient Chinese medical tradition had only two factors, yin and yang, although a kind of energy called qi was also highly important. In all three traditions, people were thought to get sick when the factors were out of balance with one another, but remedies could restore balance and make people healthy again. Mayan and Aztec medicine also revolved around balance metaphors.

Greek Humor Theory

Hippocrates was a Greek physician born around 460 BC who was the first Western thinker to attribute diseases to natural rather than theological causes. The essay *On the Sacred Disease*, written by him or by his followers, described the symptoms of epilepsy and attributed them to the accumulation of phlegm in the veins of the head rather than to the actions of the gods. More generally, another treatise attributed to Hippocrates set out the humoral theory of disease: "The body of man has in itself blood, phlegm, yellow bile and black bile; these are the things that make up the nature of his body and through these he feels pain or enjoys health. Now he enjoys the most perfect health when these elements are duly proportioned to one another in respect of compounding, power, and bulk, and when they are perfectly mingled. Pain is felt when one of these elements is in defect or excess, or is isolated in the body without being compounded with all the others."[2] Balance means having the right proportions of the key factors. *Proportion* is also a metaphor, borrowed from the mathematics of fractions. Another metaphor commonly used for balanced humors is *harmony*, borrowed from music.

The four humors are bodily fluids that only approximate currently known substances. The Greek concept of blood was much the same as today, although blood was thought to be manufactured in the liver rather than the bone marrow, which is the origin of most blood cells. Today we think of phlegm as the mucus that we cough up when we have a cold, but the Greek view included a wider range of fluids that included saliva, plasma, and lymph. Today, bile is a dark green or yellowish-brown fluid produced by the liver that aids digestion of fats in the small intestine. The Greeks distinguished two kinds of bile, yellow and black, which do not map onto current medical categories. The old-fashioned word for psychological depression, "melancholy," derives from Greek words for black and bile. Perhaps the Greeks' yellow bile corresponds to the current concept of bile and black bile is a manifestation of clotted blood.

The four humors are associated with the four elements that the Greeks thought were the basis for all matter: blood with air, phlegm with water, yellow bile with fire, and black bile with earth. Today, these four elements have been superseded by the periodic table, with 118 elements that are far more fundamental than the Greek four. Air is a mixture of gases such as oxygen and nitrogen, water is a compound of hydrogen and oxygen, fire is a process of rapid oxygenation, and earth is a mixture of various elements such as carbon and silicon.

The balance of humors in a person was thought to determine personality: a sanguine person had a great deal of blood, which produced energy and enthusiasm; a phlegmatic person had too much phlegm, which produced lethargy; a choleric person had too much yellow bile, which induced irritability; and a melancholic person had too much black bile, which led to social withdrawal. Similarly, health and disease depended on humoral balance, with vertigo resulting from an excess of black bile and fevers resulting from too much yellow bile.

It is not easy to specify the source concept for the Hippocratic metaphor of balance. The two main candidates are bodily

TABLE 7.1 Analogy underlying the metaphor of balanced humors

Weight scale (source)	**Humors (target)**
Two pans	Four humors
Weight on each pan	Amount of each humor
Balanced when equal	Balanced when proportionate
Unbalanced when unequal	Unbalanced when disproportionate
Correct imbalance by increasing and decreasing weights	Restore health by increasing and decreasing humors

balance and the weight scale, which the Greeks knew well. The Greek goddess of justice, Themis, was sometimes depicted as holding a weight scale. Table 7.1 outlines a possible mapping between the weight scale and humors, complicated by the fact that two pans correspond to four humors, so the mapping is not one to one. Nevertheless, the basic idea comes through that it is good when the weights are equally balanced and similarly good when the humors are balanced. However, humors are multidimensional, in contrast to the single dimension of weight in the balance scale.

Nothing bad happens when the weight scale is unbalanced, but Hippocratic theory says that when the humors are unbalanced the result is disease. Accordingly, diseases can be treated by correcting the imbalance: medicine is just subtracting what is in excess and adding what is wanting. In this analogy, the target is the process of disease resulting from humors, and the source is the visual and tactile process of adjusting weight scales. The correspondence is dynamic because it requires repeated adjustments and it compares causal relations of increase and decrease. Because the balance scale predated the humoral theory of disease by hundreds of years, the balance comparison was probably conventional rather than novel.

The main method of restoring balance is diet, because different foods produce different amounts of the four humors, but environmental changes in heat, cold, and wind can also help. In addition, humor levels can be directly changed by bloodletting, which removes an excess of blood. Bloodletting remained a standard method of Western medicine well into the nineteenth century, along with other ways of adjusting the humors, such as induced vomiting. Health is emotionally positive, carrying over the usual desirability of physical balance, as shown in figure 7.1.

The humoral theory of disease was amplified by the Roman physician Galen and held sway in European and American medicine until Pasteur developed the germ theory of disease in the 1860s. Modern Western medicine has developed mechanistic explanations for many kinds of illness, including ones that are infectious (COVID-19), metabolic (diabetes), cardiovascular (hypertension), nutritional (scurvy), and genetic (Huntington's). Treatments optimally consist of changing the underlying causation, for example, by using drugs to kill infectious agents such as

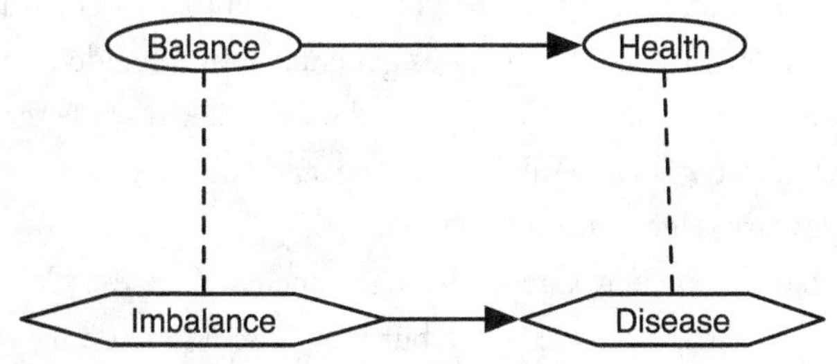

FIGURE 7.1 Transfer of emotion from bodily balance to health. Ovals indicate positive values, while hexagons indicate negative values. Dotted lines indicate incompatibility, and lines with arrows indicate emotional transfer from source to target.

bacteria and viruses. Today, *balancing* is no longer a dominant metaphor in Western medicine except for balance disorders such as vertigo and balance ideas related to bodily systems, discussed later.

The *humoral balance* metaphor was intended to explain why people get sick and to solve the problem of how to treat diseases. Even in its day, the metaphor was weak because the mapping between humors and scales or body balance was imprecise and lacking in causal underpinnings. Today, the metaphor is easily assessed as bogus and escapes qualifying as toxic only because the humoral theory is rarely used except in obscure variants of alternative medicine.

Traditional Chinese Medicine

The *Yellow Emperor's Classic of Internal Medicine* is a standard source of traditional Chinese medicine that was probably written a few hundred years BC, after the time of Hippocrates, although many of its ideas are much older.[3] It took from Taoist philosophy the idea that nature consists of two aspects that are both opposite and complementary. These aspects are called yin and yang, which were originally the dark and light side of a mountain. The opposition was generalized to include that the moon is yin while the sun is yang, the earth is yin while the sky is yang, water is yin while fire is yang, and cold elements are yin while warm elements are yang.

Like the ancient Greek elements, ancient Chinese elements included earth, fire, and water, but they differed in adding wood and metal but not air. These elements are strongly associated with yin and yang, with hot elements such as fire being yang and cold elements such as water being yin. In the universe, yin and yang are in balance, as suggested by the standard image in figure 7.2.

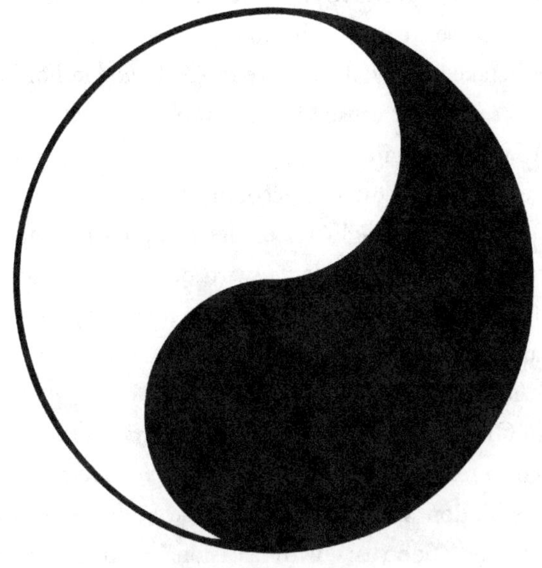

FIGURE 7.2 The symbol for yin and yang showing a unity of opposites.

Source: Wikimedia Commons.

In general, yin and yang can be transformed into each other, and such transformations occur in the human body. Functional activities in the human body (yang) are supported by the body fluids (yin), producing nutrients (yin) that support the functional activities (yang). Conversely, functional failure of the internal organs (yang) jeopardizes the digestion of food into usable nutrients (yin). The equilibrium of yin and yang produces energy (qi) to maintain the physiological functions of the body.

Accordingly, all diseases result from an imbalance of yin and yang, with too much yin bringing about cold and too much yang bringing about heat. As with the Greek humors, the factors yin and yang can occur with excesses and deficiencies, which cause disease. Unlike the Greek humors, which are recognizable fluids, yin and yang are abstract principles. But the causes and treatments of disease are similar: if you have too much of a contributing factor, take it away; if you have too little, add it. For example, vertigo is said to be caused by liver problems that result in an excess of yang over yin.

Table 7.2 shows how this account of disease corresponds to the operation of a weight scale. The mapping is simpler than the one for humors in table 7.1, with only two balance factors instead of four. Like the humor balance metaphor, the underlying analogy is dynamic and relational. Balance is clearly desirable because it brings health.

As in Greek medicine, the Chinese treatments for disease are intended to restore balance through dietary changes and physical manipulation. Hundreds of Chinese herbal products have been used for 2,000 years, with occasional medical significance. The 2015 Nobel Prize in physiology or medicine was awarded to Youyou Tu from the Academy of Traditional Chinese Medicine, whose group used traditional recipes for treating fever to discover a plant that inhibits malaria parasites. But the following core claims have received no empirical validation: the existence

TABLE 7.2 Analogy underlying the metaphor of balanced yin and yang

Weight scale (source)	**Yin-yang (target)**
Two pans	Yin and yang
Weight on each pan	Amount of yin and yang
Balanced when equal	Balanced when proportionate
Unbalanced when unequal	Unbalanced when disproportionate
Correct imbalance by increasing and decreasing weights	Restore health by increasing and decreasing yin and yang

of yin, yang, and qi; the causal efficacy of imbalances of yin and yang to produce disease; and the effects of herbal products in changing such imbalances.[4]

Whereas ancient Greek medicine used bloodletting and purging as methods of physical manipulation to change humoral balance, traditional Chinese medicine's main manipulation is acupuncture. Acupuncture is the insertion of needles to improve the flow of qi in the body along its meridians. Qi is a kind of vital energy, and meridians are channels that are different from blood vessels or nerves. Qi is responsible for the functions and operations of the body's organs and depends on a balance between yin and yang. Blockages of qi cause diseases such as heart problems that can treated by using acupuncture to restore the flow of qi through the body.

The current status of acupuncture in scientific medicine is unclear.[5] Critics of acupuncture argue that the studies that purport to show that it is effective for problems such as pain and nausea are poorly done and have inadequate controls. The apparent successes and the continuing popularity of acupuncture can be explained away as placebo effects. Defenders of acupuncture point to increasing numbers of randomized clinical trials, some of which blind participants by using sham acupuncture that stimulates nonmeridian regions. The success of acupuncture on animals undermines the placebo-effect explanation.

A mechanistic explanation of the operation of acupuncture finds support for the claim that acupuncture stimulates nerves that activate endorphins and other painkillers in the brain. Support for this mechanism comes from findings that giving people the drug naloxone, which blocks opioid function in the brain, also blocks the pain-killing effects of acupuncture. This neurological mechanism is radically different from the metaphor of balancing qi, yin, and yang. Another physical way of manipulating qi, yin, and yang is a practice involving slow movements and breathing called qiqong, a variant of which is the exercise form tai chi, which is examined later.

Traditional Chinese medicine is still widely practiced in China and Taiwan alongside Western medicine. In North America and elsewhere, it is sometimes used as a complement to Western medicine. Its methods, including herbal treatments, acupuncture, and tai chi, are being investigated by the U.S. National Institutes of Health through a well-funded National Center for Complementary and Integrative Health. Evidence for the clinical efficacy of such treatments remains scant, and I argue below that traditional Chinese medicine is a pseudoscience. The claim that disease results from imbalances of yin, yang, and qi is as shaky as the claim that disease results from humoral imbalances. Because the Chinese theory is far more popular today than the humoral theory, it is more dangerous because it can draw people to practices that are much less effective than evidence-based treatments in mainstream medicine, such as using inert herbs to treat potentially lethal cancers. Therefore, the balance metaphor at the core of traditional Chinese medicine should be rated as not only bogus but also toxic.

Indian Ayurvedic Medicine

The Indian counterpart of traditional Chinese medicine is Ayurveda, a term from Sanskrit words for "knowledge of life."[6] The traditions of Ayurveda go back thousands of years but were only written down a few hundred years BC, around the same time that Greek humor theory and traditional Chinese medicine were codified. Ayurvedic medicine is still widely practiced in India and also has followers in North America through the influence of best-selling author Deepak Chopra, who combines

Ayurvedic practices with New Age mysticism and a simplistic and distorted version of quantum physics.

Most generally, Ayurveda aims for a balance between mind and body, but specifically it aims for balance among three doshas, which are elemental forms of energy: vata, pitta, and kapha. Health is maintained when these three are balanced, and imbalance results in diseases. Vata is a kind of energy associated with cold, light, and movement, and unbalanced vata can lead to diseases such as gout. Pitta is associated with heat and moisture, and with fever when it is unbalanced. Kapha combines earth and water to produce nourishment, and imbalance can lead to diseases such as bronchitis.

The three doshas are associated with the five classic elements of ancient India: air, fire, water, earth, and ether. Ether is what exists beyond the material world and is different from the fifth element that Aristotle added to the previous four to explain the movements of the planets and stars. Vatha is most closely associated with air, pitta with fire, and kapha with earth.

Different people have different personal characteristics depending on the preponderance of doshas in their bodies, just as personality for the Greeks depended on the preponderance of humors. People with vata personalities are quick thinking and fast moving, pitta personalities are fiery, and kapha types are calm. Medical treatments to restore balance among the doshas need to take into account these personalities, with different kinds of people needing different diets. Like yin and yang but unlike the bodily humors, the doshas are abstract entities that are not directly observable, adding transbodiment to the embodied sense of balancing. Particular maladies result from imbalances, as when vertigo arises from excessive pitta.

As for the Greek and Chinese balance metaphors, the Ayurvedic source analog is more plausibly the weight scale

TABLE 7.3 Analogy underlying the metaphor of balanced doshas

Weight scale (source)	**Doshas (target)**
Two pans	Vata, pitta, and kapha
Weight on each pan	Amount of vata, pitta, and kapha
Balanced when equal	Balanced when proportionate
Unbalanced when unequal	Unbalanced when disproportionate
Correct imbalance by increasing and decreasing weights	Restore health by increasing and decreasing vata, pitta, and kapha

rather than bodily balance, as laid out in table 7.3. Some versions of Ayurveda require the three doshas to be exactly equal, although other versions view proportions as more complex and depending on individual differences. The target analog of doshas is relational, dynamic, and multidimensional, while the weight scale is a conventional source. Balance is healthy and therefore desirable, like the achievement of equality in the weight scale and especially the maintenance of bodily balance.

As with traditional Chinese medicine, diet is the major way of treating diseases by restoring balance, but there is less emphasis on herbal remedies. Meat is discouraged for all people, in accord with the traditional vegetarianism of Hinduism. Spices such as cardamom and cinnamon can help to balance the doshas, along with hundreds of roots, fruits, and seeds.

Ayurveda does not have physical manipulations like humoral bloodletting or Chinese acupuncture. But it does recommend exercise to adjust the doshas, and some yoga poses work specifically on particular imbalances, as when people with too much pitta perform abdominal twists. Yoga in Ayurvedic medicine has some of the same goals as qigong and tai chi in traditional

Chinese medicine, and I pursue the comparison below. Ancient Greek medicine also recommended exercise but more in the form of sport than in meditative breathing.

Like the Greek and Chinese balance metaphors, the Ayurvedic balance metaphor is supposed to provide medical explanations and treatments, but no evidence supports claims that it accomplishes these purposes. Its contemporary popularity is explained in India by tradition and elsewhere by the alluring gibberish of writers like Deepak Chopra. Sadly, modern scientific medicine cannot deal with the full range of people's medical problems, leaving them highly motivated to find solace in flowery promises of easy fixes to health problems through eating the right foods and practicing yoga. As long as people only use Ayurvedic medicine to supplement evidence-based medicine, the bogus nature of its underlying metaphor is not harmful, but the metaphor turns toxic if it is used to supplant effective treatments.

EVALUATING MEDICAL METAPHORS

Balance metaphors are highly appealing. People are familiar with balance from experience with their bodies and with weight scales. Balance is obviously good, so it makes sense when we are told to balance our lives, health, and diet. The identification of health with balance was natural for ancient Greek, Chinese, and Indian physicians and their patients, and it remains natural for current advocates of alternative and complementary medicine.

Whereas the Greek humors correspond to the real substances of blood, phlegm, and bile, the three doshas are hypothetical substances for which there is no scientific evidence. Similarly, yin and yang are hypothetical entities derived metaphorically from the dark and light sides of the mountain. What is common to the three ancient medical traditions is that they all employ the metaphors that health is balance and disease is imbalance. It does not matter whether the substances allegedly in balance are real (humors), hypothetical (doshas), or metaphorical (yin/yang): the balance metaphor works the same in all cases. Balance metaphors are usually applied to real phenomena, as in the COVID-19 predicament of balancing lives and livelihoods, both of which are literal. The balanced elements do not have to be metaphorical for the overall expression to be metaphorical, because *balance* and *imbalance* are the metaphors.

For poetic purposes, metaphors only need to be appealing, but much more rigid standards are required for ones that are supposed to provide practical guidance. Some metaphors are highly misleading, as I showed for the balance of nature. What standards can we use to distinguish insightful metaphors from misleading ones? The key question is whether the claims made in a medical metaphor are true, explanatory, and curative.

Evidence

Here are some specific questions to ask about metaphor-based theories in medicine.

1. Do the entities proposed in the theory exist?
2. Do the entities interact in ways that can produce health and disease?
3. What evidence is there that balance problems cause disease?
4. What evidence is there that restoring balance heals diseases?
5. What evidence is there that maintaining balance promotes health?

All three ancient medical balance theories score poorly on these questions.

On the first question about the existence of entities, the Greek humor theory does better than the Chinese and Indian theories. Blood, phlegm, and some version of bile are easily observed, but no observations support the existence of yin, yang, qi, vata, kapha, and pitta. Of course, science routinely forms hypotheses about nonobservable entities such as atoms, electrons, quarks, atomic bonds, genes, and mental representations. But these hypotheses are rigorously evaluated as to whether they provide better explanations than competing hypotheses of the full range of evidence. The three balance theories of medicine fare poorly in explanatory power compared to the Western medical theories based on cellular and molecular mechanisms that apply to infectious, cardiovascular, genetic, nutritional, metabolic, autoimmune, cancerous, and other diseases.[7]

The elements allegedly associated with the balance factors in ancient theories also have evidence problems. According to current physical theory, the 118 known elements do not include earth, air, fire, water, metal, wood, or ether. The last-mentioned does not exist, and the first five can all be understood as mixtures, compounds, or processes resulting from mechanisms involving the 118 elements. Moreover, no evidence supports the existence of qi and doshas as forms of energy. In contrast, the modern physical theory of energy as the capacity to move force through a distance relates to well-documented forms of energy, including electrical, mechanical, and atomic energy, which are subject to well-established laws such as the conservation of energy. We have ample reason to believe in these forms of energy, but no evidence that supports fuzzy concepts like qi and doshas.

Contrast the neurophysiological explanations of vertigo in chapter 3 with the explanations provided by the three ancient theories. Various vertigo disorders, such as the benign positional version, result from interactions of neural mechanisms operating in the inner ear, brain, and body. The relevant parts include the hair cells in the inner ear and the neural groups in the vestibular nuclei. The existence and interactions of these parts are well established through experimental investigations by many different scientists, with general consensus in the scientific community. Some treatments consistent with these mechanisms have proven effective, such as the Epley maneuver for benign positional vertigo. In contrast, the ancient claims that vertigo results from imbalances due to excesses of substances such as black bile, yang, and pitta are just guesses that have had no empirical evaluation of either their accuracy or treatment success.

Good evidence has these characteristics:[8]

1. *Reliability.* A source of evidence is reliable if it tends to yield truths rather than falsehoods, as in systematic observations using instruments such as telescopes and microscopes and in controlled experiments such as those practiced by many scientists. In contrast, religious texts, hallucinations, gossip, and philosophical thought experiments often support falsehoods.
2. *Intersubjectivity.* Systematic observations and controlled experiments do not depend on what any one individual says, but are intersubjective in that different people can easily make the same observations and experiments.
3. *Repeatability.* Systematic observations and controlled experiments have intersubjectivity because the same person or different persons can get similar results at different times, replicating the original experiments or observations.
4. *Robustness.* Experimental results should be obtainable in different ways, such as by using different kinds of instruments

and methods. For example, different kinds of microscopes can be used to provide similar insights into cell structure.

5. *Causal correlation with the world.* Evidence based on systematic observation or controlled experiments is causally connected with the world about which it is supposed to tell us. For example, telescopes and microscopes provide evidence because reflected light enters the eyes of observers, stimulates their retinas, and generates perceptions in accord with neural processes that are causally regular.

In medical research, the most reliable evidence comes from randomized clinical trials. If you want to know if a herb such as ginseng is helpful for brain function, you cannot just give it to a few people and ask them if they feel better. Instead, you need to get a large sample of people and randomly divide them into two groups, one with people who get ginseng and the other who get something else not related to brain function. The experiment should be blinded in that people do not know if they are getting ginseng or the alternative. Ideally, the experimenters should also not know which people are getting ginseng so that their biases cannot affect how they determine its effectiveness. It is also important to have an objective measure of brain function, such as performance on memory tests.

If these conditions are satisfied and the people in the ginseng group show improved brain function, then we have good evidence for the effectiveness of the treatment, subject to further experimental tests. When done carefully, randomized, blinded clinical trials provide evidence of efficacy that fits with the five standards of evidence. Such trials do not completely eliminate alternative explanations for their observations, such as chance, experimenter bias, or fraud, but they do provide reason to believe that the best explanation of the trials' result is that the treatment involved is genuinely effective.

The Cochrane Collaboration is an international not-for-profit organization that prepares and maintains systematic reviews of randomized trials of health-care therapies. A 2009 article reviewed seventy studies of traditional Chinese medicine treatments such as herbs and acupuncture.[9] Although the authors identified a few trials suggesting that some treatments might be effective, their overall conclusion was that there is no conclusive evidence that traditional Chinese medicine is beneficial. Similarly, Cochrane evaluations of Ayurvedic treatments for schizophrenia, diabetes, and irritable bowel syndrome found insufficient evidence to recommend their use.

Ancient Greek, Chinese, and Indian medicine all contain claims about the causes and treatments of disease that have not been supported by clinical trials. The lack of such trials is not the only reason for doubting these approaches, according to criteria commonly used by epidemiologists.[10] In determining whether a behavior causes a disease, such as whether smoking causes cancer, it is important to look for the biological mechanisms by which the behavior could cause the disease. In the 1960s, the hypothesis that smoking causes cancer was mostly supported by experimental studies, but well-established mechanisms are now known: cigarette smoke contains chemicals that damage the DNA in cells, leading to genetic mutations that produce cancer.

Another criterion used by epidemiologists is that medical claims should be coherent with established scientific knowledge. What is established changes over time as new evidence and theories lead to sometimes revolutionary developments in science. But a huge amount of evidence supports current theories in physics, chemistry, and biology such as relativity, quantum theory, atomic theory, and genetics. So we should expect some degree of fit or at least consistency between medical theories and these theories.

General balance theories such as traditional Chinese medicine and Ayurvedic medicine fare poorly on the criteria of mechanism and coherence. As with the four humors, no mechanisms are known by which unbalanced yin, yang, qi, vata, kapha, and pitta cause disease or mechanisms by which rebalancing them can restore health. Moreover, we should not expect to find such mechanisms because the operation of the alleged balance factors is incoherent with the array of causal mechanisms in current natural science for which ample evidence exists. The ancient balance theories of China and India should be as medically defunct as the Greek humoral theory, which faded away in the nineteenth century.

Pseudoscience

A pseudoscience is a field that claims to be scientific but is not.[11] The humoral theory today survives as a practice of some naturopaths who try to pass it off as scientific. Similarly, traditional Chinese and Ayurvedic medicine sometimes claim to be alternatives or complementary to Western medicine. These approaches can be assessed using the following differences between science and pseudoscience that I developed from reflections on astrology and other dubious practices.

1. Science explains using mechanisms, whereas pseudoscience lacks mechanistic explanations.
2. Science uses correlation thinking, which applies statistical methods to find patterns in nature, whereas pseudoscience uses

dogmatic assertions, or resemblance thinking, which infers that things are causally related merely because they are similar.

3. Practitioners of science care about evaluating theories in relation to alternative ones, whereas practitioners of pseudoscience are oblivious to alternative theories.
4. Science uses simple theories that have broad explanatory power, whereas pseudoscience uses theories that require extra hypotheses for particular explanations.
5. Science progresses over time by developing new theories that explain newly discovered facts, whereas pseudoscience is stagnant in doctrine and applications.

In line with the first difference, humoral, traditional Chinese, and Ayurvedic medicine lack mechanisms and shun them in favor of being "integrative" or "holistic." They are clearly dogmatic, asserting ancient wisdom about balance rather than indicating why it was wise. Resemblance thinking from similarities to causal claims is sometimes used, as when traditional Chinese medicine proposes that plants resembling parts of the body can be used to treat ailments of those parts. Evaluation in comparison with alternative theories is rare—I have found no competitive assessment of Chinese herbal treatments versus Ayurvedic ones. Only in recent years through agencies such as the NIH National Center for Complementary and Integrative Health have comparisons been made with Western treatments, although acupuncture has sometimes received experimental evaluation.

My fourth difference between science and pseudoscience does not count against the yin/yang/qi and three-dosha balance theories because these employ a small number of hypotheses. But the fifth difference is striking because humoral, traditional Chinese, and Indian medicine are remarkably stagnant, bragging about their maintenance of principles and practices that are thousands of years old. In contrast, modern medicine took off in the middle of the nineteenth century and has made dramatic progress in explaining infectious, genetic, metabolic, autoimmune, and nutritional diseases. COVID-19 was only identified as a disease in January 2020, but within a year research brought breakthroughs in identifying the responsible coronavirus, determining its genetic structure, establishing tests for the disease, treating the disease using antiviral drugs and anti-inflammatories, and generating effective vaccines. A few reports describe people with COVID-19 being treated with Chinese herbs and Ayurvedic tablets, but without scientific evaluation.

Popularity

On four of the five differences between science and pseudoscience, the balance theories in humoral, traditional Chinese, and Ayurvedic medicine qualify as pseudoscience. Then why are they so popular in their countries of origin and in Western circles that value alternative and complementary approaches? The reasons for their success are biological, psychological, and sociological.

The biological reason for this enduring popularity is that modern Western medicine, despite its accomplishments, has shortcomings. There are many diseases for which knowledge of the underlying mechanisms of organs, cells, and molecules is insufficient to provide full explanations and treatments. Of the vertigo diseases described in chapter 3, benign positional paroxysmal vertigo is the only one that has both a full mechanistic explanation (errant crystals in the inner ear canals) and effective treatment (the Epley maneuver). Much is known about the mechanism breakdowns that lead to Ménière's disease and brain-related vertigo, but full explanations of their origins are still under investigation and treatments are limited.

Hundreds of other diseases are missing biological knowledge sufficient for understanding their causes and cures. The antiviral drugs currently used to attack the novel coronavirus only slow its effects, and vaccines are never 100 percent effective. Common diseases such as cancer, high blood pressure, and diabetes have available treatments, but millions of people die of them every year. The limitations of modern Western medicine drive people to look elsewhere for help with their health problems. It is also natural to be skeptical of claims by pharmaceutical companies that maximize their profits by marketing rather than developing new and effective treatments.

The second set of reasons that balance theories of medicine remain popular is psychological. Balance explanations make sense to people because they are already familiar with the importance of bodily balance and the usefulness of weight scales. Moreover, people are highly motivated to believe that treatments from traditional Chinese or Ayurvedic medicine will work for them because they want to be healthy. Psychological research has documented the ubiquity of motivated inference, which distorts thinking when people reach conclusions based on their goals rather than on reliable evidence.[12] Motivated inference is more complicated than wishful thinking because people do not just believe what they want to. Instead, they are selective in remembering and collecting evidence in ways that serve their goals. For example, someone who wants to survive cancer will eagerly look for untested treatments that promise a cure rather than carefully assess the evidence for their effectiveness.

Another psychological reason for the popularity of bogus treatments is that they often work because of placebo effects.[13] Just thinking that a drug might work can actually make it work, as we know from countless drug trials that have to control for placebos. Giving people an inert substance can make them feel better because of complex interactions between the brain and the rest of the body.

Finally, the causes of the continuing popularity of ancient balance theories of medicine are partly sociological. Modern China and India each have more than a billion people accustomed to using their familiar medical treatments for everyday ailments. The governments of these countries have little political motivation to replace traditional treatments with Western ones that are much more expensive to deliver. In the West, people hear about alternative treatments from friends and from social media sources like Facebook that are virulent sources of medical misinformation such as hostility toward vaccines. Another social factor is the populist trend to distrust experts as part of the elite.

These biological, psychological, and sociological reasons explain why pseudoscientific alternative medicine survives and thrives. This influence is relatively harmless as long as it does not prevent people from getting effective treatments for their medical problems, but acute dangers can arise. For example, some people are so averse to vaccines based on misinformation about harmful side effects that they refuse COVID-19 vaccines, whose effectiveness is empirically validated. The resulting noncompliance could cause countless additional cases and deaths.

TAI CHI AND YOGA

All three ancient medical traditions sensibly recommend eating well and exercising as keys to good health. Today abundant evidence supports claims that exercise has great benefits for physical and mental well-being by virtue of biological mechanisms.[14]

In contrast, the ancient Greek, Chinese, and Indian traditions explained the value of exercise in terms of its contribution to balancing humors, qi, or doshas.

The Greeks gave us Olympic sports, and the other two traditions provided unique exercise practices that are still used today, tai chi and yoga. Evidence finds health benefits for both of them that is independent of their alleged influence on internal imbalances. Moreover, both tai chi and yoga help people improve their bodily balance and avoid falls, which are a major threat for old people. I see no reason to attribute their success to adjustments of yin/yang or the three doshas but instead explain their success in improving bodily balance based on the neural mechanisms presented in chapters 2–4.

Why Does Tai Chi Feel Good?

Tai chi is an ancient Chinese tradition that provides a gentle and graceful form of exercise. I took a course in 2019, but my second course was interrupted by the COVID-19 lockdown. Thinking about how tai chi helps with balance inspired the book you are now reading.

The evidence-based health benefits of tai chi are substantial for helping with falls, osteoarthritis, Parkinson's disease, chronic obstructive pulmonary disease, cognitive incapacity in older adults, depression, dementia, and cardiac and stroke rehabilitation.[15] Moreover, weaker evidence finds improvements in psychological well-being, including reduction of stress and anxiety and increased self-esteem.

What explains these benefits? Traditional Chinese medicine says that tai chi works by balancing yin and yang and redistributing qi energy. Psychology offers a different set of explanations of why tai chi works through cognitive, emotional, and social mechanisms.

Cognitively, tai chi is more complicated than it looks. The slow movements in tai chi seem simple but actually take a lot of concentration. You need to keep track of both arms, both legs, and hips, with novel movements such as forming a hook with a hand. Moving slowly requires more concentration than a faster, jerkier movement. In addition, the movements are accompanied by controlled deep breathing in when your hands go up or toward the body, and breathing out when hands go down or away from the body.

Moreover, a short, minute-long sequence in tai chi can require ten different combinations of movements, each of them with four different movements of hands and feet. So a five-minute sequence can require around 200 different actions, not including controlled breathing. Thus tai chi imposes a large cognitive load on the mind.

This cognitive load prevents your mind from wandering. I have tried meditation but never have been able to do it for more than a few minutes because it is too boring and my mind wanders to more interesting aspects of my life. In contrast, tai chi requires full concentration, reinforced by my teacher, who noticed if I was not following precise instructions. Concentrating on body movements prevents people from thinking about other, more stressful aspects of their life such as work, health, and family conflicts.

How this prevention works is explained by the theory of consciousness presented in chapter 4. Your brain forms mental representations that are different patterns of firing in large groups of neurons. These representations compete with each other for the limited span of consciousness—you can only keep around five to seven things in mind at once. The complex movements of tai chi require new kinds of motor representations that take over consciousness, outcompeting troubling thought. I think this is one of the reasons that tai chi reduces stress.

Stress is also tied to emotions, and tai chi has emotional effects that are more than just cognitive competition. Emotions depend both on cognitive appraisals of how a situation is affecting your goals and on detection of physiological changes, where appraisals and changes are represented by unified neural representations. Tai chi does not raise the heart rate like more vigorous exercise, but the deep breathing definitely affects physiology in the way that meditation does, producing a calming effect.

Deep breathing moderates the vagus nerve, which is the longest part of the autonomous nervous system, connecting the heart, lungs, and digestive track. Calming the body sends signals to the brain that complement the reappraisal that comes from not being able to think about stressful aspects of life. So tai chi lowers stress by regulating emotions as well as by diverting thoughts.

I miss the tai chi classes because they also have important social effects. I practice tai chi on my own for about twenty minutes twice a week, but the classes are group events, as are park gatherings where large numbers of people participate. The sociologist Randall Collins emphasized the importance of interaction rituals in which mutually focused emotions and attention produce a shared reality that generates solidarity.[16] Interaction rituals are important in religious observances, sports events, dances, and live concerts. Tai chi similarly generates emotional energy from group practices, complementing the individual cognitive and emotional effects on stress reduction.

Why Do Tai Chi and Yoga Improve Balance?

Yoga is a form of exercise that originated in India around the same time as Ayurveda.[17] It uses a series of colorfully named poses, mostly performed on a yoga mat, such as the Downward Dog, where people place their hands and feet on the mat and raise their hips as high as they can. Such poses stretch and strengthen muscles and can help with balance. For example, the Tree pose requires balancing on one leg. The evidence for yoga's contribution to good balance and avoidance of falls is not as strong as the evidence for tai chi's contribution, but a few trials are at least suggestive. Yoga meshes with Ayurveda with claims that some poses help to balance vata, pitta, and kapha.

The *Harvard Medical School Guide to Tai Chi* provides numerous movements and reviews studies that support claims that tai chi improves people's balance. But in explaining why tai chi works, it falls back on the traditional explanations based on yin, yang, and qi. The neuroscience-based explanation of balance in chapter 2 can do better.

Good balance and the avoidance of dizziness, vertigo, and falls require coordination among biological systems in the inner ear, the brain, the eyes, and the body. Balancing the body is far more complicated than keeping track of weights in the balance scale because the brain has to integrate information from the movement of body parts such as the arms and legs; from different parts of the inner ear such as the canals and the otoliths; from the eyes, including both what they are seeing and how they are moving; and also from different parts of the brain, including cognitive expectations.

Systematic exercises such as tai chi and yoga train the brain and body to carry out such integration more effectively. Learning in the brain changes connections between neurons that consist of synapses. When two neurons that are connected by a synapse fire at the same time, the synaptic connection between them is strengthened so that they become more likely to fire together in the future. This strengthening is called Hebbian learning after the Canadian neuroscientist Donald Hebb, who proposed it in the 1940s. The frequently repeated motions in tai chi and yoga are well suited to produce this kind of learning by strengthening synaptic connections. The relevant connections can be fairly local in the vestibular nuclei, but also between these neurons and parts of the brain such as the cerebellum and the visual cortex.

Neural connections can also be strengthened by another kind of learning through reinforcement. Learning and practicing tai chi and yoga pose a series of challenges concerning making novel moves and repeating them with increasing skill. Meeting these challenges brings external rewards from approving teachers and also internal rewards through the personal satisfaction that comes from accomplishing difficult tasks. In the brain, these rewards are manifest as dopamine signals that change the synaptic connections between neurons whose firing led to the accomplishment of goals. Thus people get better at balancing their bodies because Hebbian and reinforcement learning modify synaptic connections that coordinate sensory and motor reactions to changing environments. Brain scanning has revealed that tai chi brings large-scale changes to the brain, including increases in brain volume, cortical thickness, and synchronization of functional brain activity.

So which is more plausible, the hypothesis that tai chi improves balance because of neural learning or that it improves balance because it adjusts yin, yang, and qi? No direct studies have shown that tai chi changes neural connections by Hebbian and reinforcement learning, but we have substantial evidence, described in chapter 2, about how balance is controlled by the brain, and also a huge amount of evidence that Hebbian and reinforcement learning are important in the brain. So well-established neural mechanisms naturally extend to explain how repeated tai chi exercise can improve balance. The same explanation applies to yoga and other kinds of balance exercises that have proved effective in decreasing falls. Accordingly, we can appreciate the health benefits of tai chi and yoga without buying into the ancient explanations in terms of yin, yang, qi, vata, pitta, and kapha. Moreover, we can look to other kinds of movement, such as dance, judo, and walking on rough terrain, that also improve balance through neural learning.

MODERN BALANCE METAPHORS IN MEDICINE

In contrast to ancient balance theories of health, modern balance metaphors are sometimes medically useful when they point to practices that are justified by experimental evidence and knowledge of underlying mechanisms. Good examples are balanced diets, immune systems, electrolytes, neurotransmitters, and autonomic systems.

Balanced Diet and Lifestyle

According to the United Kingdom's National Health Service (NHS), "Eating a healthy, balanced diet is an important part of maintaining good health, and can help you feel your best. This means eating a wide variety of foods in the right proportions, and consuming the right amount of food and drink to achieve and maintain a healthy body weight."[18] The NHS recommends portions of fruits, vegetables, high-fiber starches, dairy or dairy alternatives, high-protein foods such as beans and fish, and unsaturated oils. Good evidence supports each of these classes of foods as a way of avoiding diseases. Moreover, much is known about how these foods enhance health; for example, eating fruit provides vitamin C, which improves cell function and prevents scurvy.

The balance metaphor for diet is not just the weight scale, which only concerns two parts (the pans) and one dimension (weight). A balanced diet has at least the six dimensions mentioned, and the advice is to get enough of each. The balance metaphor guides people to get enough of each required ingredient and not go unbalanced by eating too much of any of them.

People are also advised to maintain an energy balance, which means not taking in more calories than they burn. More generally, a balanced lifestyle requires combining a good diet with exercise, sleep, relaxation, and social contacts. In chapter 8, the psychology section discusses how balance contributes to a meaningful life.

Balanced Electrolytes

An electrolyte is a substance that produces a conducting solution when dissolved in a solvent such as water. The human body requires a balance among essential minerals such as sodium, potassium, chloride, calcium, magnesium, phosphate, and bicarbonate.[19] Electrolytes are important for nervous system function, muscle function, and maintenance of hydration. Having levels of an electrolyte that are too high or low can cause fatigue, irregular heartbeat, confusion, and other dysfunctions. Ample evidence supports the connection between electrolyte imbalances and feeling poorly, and the chemistry of the relevant substances furnishes the underlying mechanisms. Homeostasis and the maintenance of equilibrium are useful balance metaphors for explaining the value of regulating electrolytes and other important physiological processes such as body temperature and blood sugar levels.

The balance metaphor for electrolytes is like the balanced diet metaphor in usefully advising people to make sure that they get enough—and not too much—of valuable nutrients. It amounts to an extended, multidimensional variant of the weight scale source, where too much weight on a single scale brings it out of equilibrium.

Immune System Balance

People are subject to diseases that result from infections by pathogens such as bacteria, viruses, fungi, and parasitic worms. The immune system recognizes these invaders and attacks them using white blood cells and other killer cells. These mechanisms are usually effective but sometimes can overreact and attack the body's own cells, producing autoimmune diseases such as psoriasis, lupus, rheumatoid arthritis, and multiple sclerosis. Some severe cases of COVID-19 are the result of overreaction by the immune system against the coronavirus, a loss of balance that can lead to inflammation and organ failure.

A balanced immune system is strong enough to defeat infections but not so strong that it attacks the body's own cells, such as the skin cells that go awry with psoriasis.[20] The source for this metaphor is presumably the weight scale, where we want each side to be not too light or too heavy. Similarly, we want a balanced immune system to be strong enough to attack pathogens but not so strong that it attacks the body.

Talk of balancing the immune system is good because it summarizes a large body of evidence that an overactive immune system can cause diseases such as lupus. The mechanisms by which cells in the immune system attack pathogens and normal body parts are also well known. Treatments for autoimmune diseases such as immune-suppressing drugs are intended to restore the required balance. The immune system contains no literal balancing, but the balance metaphor provides a convenient way of pointing to its proper and improper operation.

Balanced Neurotransmitters

Brain function requires more than a hundred chemicals that help to transmit signals from one neuron to another. Important neurotransmitters include dopamine, serotonin, GABA, acetylcholine, and glutamate. Too much dopamine action is a factor in schizophrenia, while a dopamine deficiency is tied to Parkinson's disease. Similarly, too much serotonin induced by LSD and other hallucinogens can lead to fearsome experiences, while too little serotonin is implicated in depression. The brain thus needs a neurochemical balance to function properly.[21]

In the 1980s, when Prozac became available for treating depression, the chemical imbalance theory of mental illness became popular. This metaphorical explanation turned out to be too simple because later work discovered other factors involved in depression, such as neurogenesis and interactions among complexes of neurochemicals. Nevertheless, the idea of chemical balance in the brain retains some usefulness because it points toward the health requirement of having approximately the right amounts of neurotransmitters to support brain functions. I conjecture that the metaphorical source for the chemical balance metaphor was the previous medical metaphor of balanced nutrition rather than the background metaphors of the weight scale and body balance. The metaphor of neurotransmitter imbalance is only slightly embodied because of its distance from scales and body balancing and because of the unobservable, transbodied nature of neurotransmitters such as dopamine and serotonin.

Neurons and other cells get their energy from components called mitochondria that undergo fission and fusion. Mitochondria malfunctions are implicated in hundreds of diseases such as diabetes and bipolar disorder, and their proper functioning depends on maintaining a balance between fission and fusion. Brain functioning also requires a balance between excitation and inhibition in neural interactions.

Autonomic Balance

My final example of a modern, scientific balance metaphor in medicine concerns the autonomic nervous system, which regulates involuntary physiological processes such as respiration, blood pressure, heart rate, and digestion.[22] This system divides into the sympathetic system, which stimulates the body's fight or flight response, and the parasympathetic system, which regulates internal bodily functions such as digestion and defecation. The sympathetic and parasympathetic systems are in opposition because the former is oriented toward quick action and the later toward slower bodily processes. For example, activity in the sympathetic system increases heart rate and cardiac contraction, while the parasympathetic system slows the heart down. Autonomic balance concerns having proportional responses of the sympathetic and parasympathetic systems, and imbalance such as too much stress can cause diseases such as heart problems.

Like the metaphors for dietary, electrolyte, immune system, and neurotransmitter balance, the autonomic balance metaphor contributes to the two major goals of medicine: disease explanation and treatment. Therefore, these five modern metaphors qualify as strong, in contrast to the bogus status of the three ancient balance metaphors. They show a place for balance metaphors in medical sensemaking, which should remain wary of bogus uses.

MEDICAL BALANCE

Darwin's brilliant metaphor of natural selection was based on the source analog of artificial selection performed by breeders. It was enormously successful in guiding development of his theory of evolution but lacked mechanisms to explain how features are passed from one generation to another and spread through a species. These gaps were eventually filled in by theories about genes, DNA, and population genetics.

Similarly, the balance metaphors concerning diet, electrolytes, the immune system, neurotransmitters, and the autonomic systems are all fleshed out by research that identifies both underlying mechanisms and correlations with disease. In contrast, the ancient medical balance traditions and their current variants have missed the mechanistic boat. They operate with loose correlations between contributing factors and diseases and between diseases and curative treatments, leaving no reason to believe that diseases result from imbalances in the four humors, three doshas, or two opposites of yin and yang. Medicine teaches valuable lessons about the perils and the promise of balance metaphors.

The worst medical balance metaphor I have seen is from a naturopathy website:

> The emphasis of natural immunity is a critical factor in the idea of Naturopathy. The body is a self-balancing system, so outside intervention would disrupt its functioning and cause the body to rely on chemically created sources for strengthening and treatment. The only external source Naturopathy approves of are supplements that are organically produced in order to boost wellness and help fight disease. Introducing foreign substances like prescription drugs, or most notably vaccines, are viewed as detrimental to the balance in the body especially to young children who have not had the time to build a sturdy immunity on their own yet.[23]

This quote is full of factual errors and encourages the avoidance of vaccines that save millions of lives. Toxic!

Balance fails as a general metaphor for health and disease but succeeds when restricted to particular bodily systems and diseases that arise from breakdowns in those systems. Other general metaphors in medicine have problems of their own; for example, thinking of a disease as a battle or war to be fought risks blaming seriously sick people as losers.[24] Perhaps the balance metaphor for health is a bit kinder than the fight metaphor, but neither is generally useful for solving the medical purposes of explanation and treatment. Medical balance metaphors need to be confined to the narrow domains where they contribute to understanding and improving illness. Similar scrutiny is required for the application of balance metaphors in social sciences.

8

SOCIETY

The eclipse of ancient balance theories of medicine means that balance metaphors are no longer central to studies of health and disease. In contrast, concepts of balance and equilibrium abound in the social realm, both in everyday discussions and in the social sciences. People strive for work-life balance, politicians struggle with checks and balances, economists explain prices in terms of equilibrium of supply and demand, and sociologists look for tipping points in racially segregated neighborhoods. In law, the central metaphor is the scales of justice, and journalists aim for balanced reporting. Bodily balance is essential for effective sports performance, and balance metaphors help to explain what makes food and wine so enjoyable.

This chapter describes the operation of balance metaphors across psychology, economics, politics, sociology, law, journalism, sports, and cuisine. But it also evaluates the dubious uses of balance metaphors, especially in economics and sociology, where the assumption that equilibrium is good can legitimize injustice. As in medicine, balance metaphors often contribute to social sensemaking but sometimes get in the way of explaining and predicting changes in people's lives.

PSYCHOLOGY

Balance metaphors about the mind are common in everyday life, but they also contribute to psychological science, where *cognitive dissonance* is the most famous balance theory. The mental health fields of psychiatry and clinical psychology strive to diagnose and treat unbalanced minds.

Everyday Psychology

In daily life, people sometimes use balance metaphors to describe personalities. Some individuals are stable and level-headed, whereas others are unstable and unbalanced. People with composure and stability have equanimity. A person suffers a fall from grace because of events that reduce status or prestige. People who are becoming too self-important can be warned: the bigger they come, the harder they fall. As with grace, falling here does not mean a literal descent in space but rather a metaphorical decrease in social standing. Similarly, a downfall is a loss of power, prosperity, or status. A vertigo metaphor applies when someone says of something confusing: "it makes my head spin."

The experience of starting to love someone is often described as falling in love, which sometimes ends with falling out of love. Falling is an imbalance metaphor that captures the lack of control that marks the beginnings and endings of intense emotional relationships. Falling in and out of love are tipping points where small quantitative changes can lead to dramatic qualitative changes, as when a trifling disagreement about cleaning the kitchen grows into a major quarrel. Using the emotional analogy shown in figure 8.1, good relationships are described as steady, settled, or stable, whereas bad ones are unsteady, rocky, or

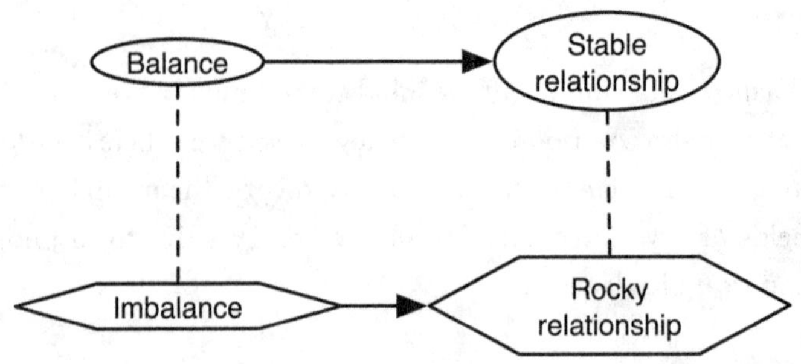

FIGURE 8.1 Transfer of emotion from bodily balance to romantic relationships. Ovals indicate positive values, while hexagons indicate negative values. Dotted lines indicate incompatibility, and lines with arrows indicate emotional transfer from source to target.

turbulent. Balanced relationships are ones with equality in areas such as communication, emotional support, and power. Individually, people seek emotional balance or equilibrium, which requires having a good combination of positive emotions such as happiness and pride with manageable amounts of negative emotions such as sadness, anger, and fear.

A major difference between psychological balance metaphors and those used for nature and medicine is that the psychological ones explicitly include the consciousness aspects of balance explained in chapter 4. Falling in love is not entirely a conscious process, but it includes conscious experiences such as feeling so excited that one is unstable, dizzy, or vertiginous—all imbalance metaphors.

Work is also described by balance and imbalance metaphors. A steady job is one where people do not need to worry about being fired or laid off. In contrast, precarious work is employment that is temporary, part-time, and on-call. Unstable work schedules produce psychological stress that can worsen mental and physical health.

The major use of balance metaphors in people's everyday lives occurs with problems of balancing love and work. Demanding jobs and family relationships come into conflicts concerning time, energy, and money. Many people struggle with work-life balance, feeling that they are falling short in both their professional accomplishments and their family responsibilities. An additional problem may be finding time for personal enjoyment and exercise. The attempt to put all these factors together is sometimes described as a balancing act, a metaphor that may have its source in circus acts such as walking tightropes. This practice gives rise to the metaphor of walking a tightrope, which is used to describe people who are having trouble balancing different aspects of their lives.

Everyday decisions are also sometimes described as balancing acts. During 2020–2022, many parents faced the difficult decision of whether to send their children back to school when there was still community spread of COVID-19. The parents had to balance the educational and social benefits of school against the health risks from the virus. Parents, teachers, and bosses all have to strike a balance between guidance and freedom for their children, students, or employees. Romantic decisions also require balancing when someone is involved with a person who has both good and bad characteristics. Work decisions such as deciding between a high-paying job and a more fulfilling one are also balancing acts. Other activities that have been described as balancing acts include communication, creativity, baking, moderation, planetary function, and art (see chapter 9).

These everyday psychology metaphors display the features presented in chapter 5. The purpose of balance metaphors in everyday life is usually explanation, when talk of falling in love, steady jobs, and work-life balance helps people to understand their own and others' predicaments. The source and target analogs are often conscious because people are familiar with feelings such as falling in and out of love, instability about work, and conflicts between life and work. The familiarity of these metaphors shows that they are conventional rather than novel.

Balancing problems in everyday life are multidimensional, making them much more complicated than the one-dimensional quality of weight scales. No single factor controls falling in love, which involves a confluence of physical attraction and more abstract appreciations and admirations. Work-life balance is intensely multidimensional because the rewards of work such as money and satisfaction of a need for achievement are different from the rewards of personal relationships such as affection and comfort. Therefore, these life balance metaphors fit better with my constraint-satisfaction account of bodily balance in chapter 2 than with the much simpler weight scale source. Table 8.1 shows the implicit analogy between body balance and work-life balance; people are usually not aware of the complex sensemaking process operating when the body has to satisfy constraints from multiple bodily sources. Achieving work-life balance became more difficult in the COVID-19 epidemic when many people had to work at home while managing family demands.

Some balance metaphors for the mind are static, such as a steady job, but others are dynamic in that they capture aspects of life that can change rapidly, such as falling in and out of love. The metaphor of work-life balance is dynamic because shifts in careers and personal relationships require constant adjustments. All of these metaphors are verbal, but they are also multisensory because vision and kinesthesia also contribute to an understanding of stability and falling. These multimodal characteristics make psychological

TABLE 8.1 Analogy underlying the metaphor of work-life balance

Body balance (source)	Work-life balance (target)
Multiple constraints from inner ear, eyes, limbs, past experience	Multiple constraints from work (money, achievement) and life (family, fun)
Strengths of constraints	Strength of constraints—emotional
Look for coherent satisfaction of constraints	Look for coherent satisfaction of constraints
Imbalance occurs when coherence is not achieved	Imbalance occurs when coherence is not achieved
Correct imbalance by changing constraints	Correct imbalance by changing constraints

balance metaphors embodied, but they are also transbodied when they introduce abstract concepts such as love and work.

Everyday balance metaphors have a strong emotional value. Falling in love can be good or bad depending on the quality of the relationship, but falling out of love usually has the negativity of body falls because of the associated pain. For work, "steady" is obviously preferred to "precarious." In balance-based decisions and life-work balance, the background assumption is that balance is good and imbalance is bad. Describing aspects of everyday life in terms of balance and imbalance enhances the appraisal of key factors in people's lives. Another emotionally negative imbalance metaphor is falling off the wagon, which applies to alcoholics who resume drinking.

What is the overall quality of these popular psychological metaphors? All appear moderately useful in describing and explaining aspects of people's lives such as personality, love, and work. They fall short of the mechanistic standards of scientific psychology, but for everyday use they have no obvious flaws or harms; so I would rate them as modestly strong. Chapter 10 looks more deeply at the question of work-life balance from a normative philosophical perspective concerned with the meaning of life.

An example of a bogus metaphor in popular psychology is the title of a book called *Leadership Vertigo*, which defines its subject in this way: "Leadership vertigo refers to false signals being sent by our brain asserting that everything is moving along as it should when it's not."[1] As chapter 3 described, in real vertigo people experience movement that is abnormal, disturbing, and contrary to how things should be perceived. Leadership vertigo is false stability rather than false movement. So the mapping between source and target in the leadership vertigo metaphor is misleading rather than helpful, which warrants its rating as bogus.

Scientific Psychology

Scientific psychology differs from popular psychology by emphasizing experiment and theory. Instead of taking for granted the concepts and assumptions that are built into ordinary language, psychology researchers try to pin them down by using empirical investigations ranging from surveys to brain scans. The evidence collected serves to evaluate theories about how minds work. These theories are usually expressed in terms of mental representations and processes, which increasingly are interpreted as neural mechanisms. Balance has a small but important role in current psychological research.

Balance metaphors for personality operate in everyday life and in ancient personality theories based on humors, yin/yang, and doshas. The dominant view of personality in scientific psychology today is the Big Five model, with the traits of openness to experience, conscientiousness, extraversion, agreeableness, and neuroticism, summed up in the acronym OCEAN. The last factor, neuroticism, is also described as emotional instability, a balance metaphor. The opposite of neurotic in this sense is being stable, confident, and resilient, another balance metaphor mentioned in chapter 6. Psychological resilience is the ability to adapt to trauma, adversity, or stress.[2] It involves different mechanisms from ecological resilience, defined in chapter 6 as the ability of a system to return to equilibrium. Both kinds of resilience indicate a capacity to move from instability to stability, and hence qualify as dynamic balance metaphors.

Clever experiments at the University of Waterloo showed a connection between physical instability and the preference for psychological stability in a potential mate.[3] Participants were randomly assigned to conditions that were either physically stable or unstable because of a wobbly table and chair. Without noticing the wobbly table, those in the unstable situation perceived less stability in other people's relationships and reported a greater desire for stability traits in a partner, for example, preferring reliability over adventurousness. The experience of physical imbalance impelled people to seek more psychological balance. Similarly, people who experienced physical instability from being in a wobbly workstation, standing on one foot, or sitting on an inflatable seat cushion perceived their romantic relationships as less stable.

Theoretically, the most important use of balance metaphors occurred in influential 1950s work in social psychology. Fritz Heider developed balance theory to explain changes in attitude that result from a drive toward consistency among beliefs and values. For example, if you like a friend who favors a particular style of music, then you will tend to like that music to maintain balance. Heider characterized a balanced state as "a harmonious state, one in which the entities comprising the situations and the feelings about them fit together without stress."[4] This description fills out the balance metaphor with three other metaphors: "harmonious" taken from music, "fit" taken from spatial shape, and "stress" taken from engineering. Other researchers tried to make balance theory more precise using graph theory but got no closer to specifying the mechanisms of psychological balance. Heider applied balance to both situations and feelings, indicating that balance concerns both beliefs and emotions.

In 1957, Leon Festinger introduced the idea of cognitive dissonance, whose popularity continues today in textbooks and popular discourse.[5] Dissonance is a distressing relation that holds between pairs of elements such as beliefs, values, attitudes, and behaviors when they do not fit together. For example, the value of staying sober is dissonant with the behavior of heavy drinking.

Like Heider, Festinger uses the spatial metaphor of fit and does little to clarify it by saying that two elements are dissonant if the obverse of one follows from the other: he fails to define "obverse" and "follows." Festinger explicitly equates his consonance with Heider's balance and his dissonance with Heider's imbalance, so it is fair to count *cognitive dissonance* as an imbalance metaphor.[6] As chapter 9 discusses with respect to music, a systematic correspondence exists between sound metaphors and balance metaphors: harmony—a simultaneous combination of tones—is a balance among tonal qualities that work well together.

Cognitive dissonance occurs when a person feels psychological stress because of conflicting beliefs, concepts, values, and actions. Festinger hypothesized that dissonance provides motivation to change the conflicting elements and used dissonance reduction to explain psychological phenomena such as decision-making, attitude change, and interpersonal behavior. For example, alcoholics may decide that sobriety is overrated, a change that Carrie Fisher dubbed wishful drinking.

Because dissonance and Heider's balance allow interactions among many elements, they do not correspond with one-dimensional weight scales and therefore can be better compared with the multiple constraints that operate on body balance, as shown in table 8.2. The mapping shows that balance and dissonance metaphors are multimodal, dynamic, transbodied as well as embodied, largely unconscious, only partly conventional, and concerned with processes. In line with the usual emotional values attached to balance and imbalance, the metaphors can be both positive (consonance) and negative (dissonance).

Festinger and Heider were vague about mechanisms for cognitive dissonance and balance, but neural networks translate the

TABLE 8.2 Analogy underlying the metaphors of cognitive balance and dissonance

Body balance and imbalance (source)	**Cognitive balance and dissonance (target)**
Multiple constraints from inner ear, eyes, limbs, past experience	Multiple constraints among beliefs, attitudes, values, and behaviors
Relations among sensory information of compatibility and incompatibility	Relations among elements of balance (consonance) and imbalance (dissonance)
Look for coherent satisfaction of constraints	Look for overall balance through coherent satisfaction of constraints
Imbalance when coherence not achieved	Distress when balance not achieved
Correct imbalance by changing constraints	Correct imbalance by changing elements

dissonance metaphor into a mechanism similar to the one proposed for balance in chapter 2. Each element, such as a belief or value, can be represented by an artificial neuron. Consonant elements can be connected by establishing excitatory links between the neurons that stand for them. Crucially, dissonant elements have inhibitory links between the neurons that stand for them. The firing rates of the neurons can be interpreted as the acceptance or rejection of the elements based on their excitatory and inhibitory links with other firing neurons.

Overall dissonance arises when it is difficult to achieve a coherent resolution of the excitatory and inhibitory connections. In this way, balance and dissonance metaphors become translated into neural mechanisms. More realistic neural networks that implement balance and dissonance can be produced when neural groups of thousands or millions of interconnected neurons, rather than individual neurons, represent beliefs or values. Metaphors of cognitive balance and dissonance deserve to be evaluated as strong because they have been plausibly translated into neural mechanisms and because dissonance explanations are still usefully applied in social psychology for purposes of explanation and prediction. Cognitive dissonance survives as a worthwhile contribution to sensemaking.

Psychological research is aimed not only at understanding minds that function well but also at finding ways to help people with mental illness. Balance metaphors contribute to pharmaceutical interventions and to understanding the success of psychotherapy. As chapter 7 mentioned, a prominent theory in psychiatry is that mental illnesses are caused by chemical imbalance. Most simply, this theory claims that depression is caused by too little serotonin and that schizophrenia is caused by too much dopamine. The theory is supported by evidence that depression is sometimes helped by serotonin reuptake inhibitors such as Prozac, and schizophrenia is sometimes helped by dopamine antagonists such as Haldol.

These imbalance metaphors are one-dimensional, with the dimension being the amount of the relevant chemical. The likely source for the metaphor of chemical imbalance is the balance scale, where too much or too little on one side of the scale will tip the balance. The chemical imbalance theory is appreciated by patients who get a simple diagnostic explanation and encouraging prognosis for their mental problems. It is also favored by pharmaceutical companies because it provides justification for doctors to prescribe their medications.

The imbalance metaphor has some value because brain functioning does require appropriate amounts and interactions of neurotransmitters such as serotonin, dopamine, norepinephrine, and GABA. But the one-dimensional metaphors are misleading for numerous reasons.[7] First, mental disorders do not always arise from the imbalance of single neurotransmitters, as seen by the efficacy of antidepressant drugs that increase levels of two or more neurotransmitters. Second, many mental disorders are most effectively treated by a combination of drugs and psychotherapy, which helps people change their cognitions and emotions. Third, mental disorders are strongly connected to social conditions such as bad relationships and economic insecurity. Fourth, mental status is partly improved by other changes in brains, such as increased generation of new neurons through increased amounts of BDNF (brain-derived neurotrophic factor), which can be achieved by exercise as well as by medications.

So the causality and treatment of mental illness are much more complicated than the chemical imbalance metaphor suggests. Accordingly, the metaphor qualifies as weak because the mapping to a one-dimensional weight scale is misleading. If excessive focus on the chemical imbalance idea leads to substandard treatment such as neglect of psychotherapy, then the metaphor deserves to be classified as toxic.

Psychotherapists can help people cope with balance problems such as tensions among personal goals, social obligations, and work demands. Emotional change can be construed as a rebalancing of different beliefs and values, in line with the balance and dissonance theories of attitudes. Psychotherapists can also help people balance their emotions, for example, by tempering periods of despair and grief with prospects of hope and happiness.

Freudian psychology suggested that the ego needs to balance the instinctual id against the critical superego, but modern psychiatry relies on neural theories rather than the ego-id-superego view of the mind. Another bogus balance metaphor operates when pop psychologists advise people to balance the left and right sides of their brains, whose organization is far more complex than the left/right distinction recognizes.

ECONOMICS

Balance metaphors pervade the economy, from balanced budgets to the concept of equilibrium, which is central in contemporary economic theory. Everyday balance metaphors largely serve their purposes, but the comparison between the budgets of individuals and the budgets of countries can be misleading. While equilibrium theories have been at the core of economic theory, the limitations of the metaphor are shown by the frequent occurrence of crises and crashes.

Everyday Economics

Balancing a checkbook or bank account means keeping track of deposits and withdrawals to ensure that the overall amount that remains—called the balance—is sufficient. This straightforward metaphor is one-dimensional because it only tracks money, so the obvious source is the weight scale: the two pans correspond to deposits and withdrawals, which count against each other. Having a balance feels good, whereas an imbalance with withdrawals greater than deposits is worrying. Businesses are obliged to balance the books to make sure that credits equal debits. Companies prepare balance sheets to report their assets and liabilities. Metaphorical and literal balance are combined in this joke: "I lost my job at the bank because an old woman asked me to check her balance so I pushed her over."

A person or family can manage finances by striving toward a balanced budget in which revenues are equal to or greater than expenditures. Failure to maintain a balanced budget can lead to financial problems such as debt, bankruptcy, and destitution. Once again, balance is good and imbalance is bad.

Problems arise when the balanced budget metaphor is extended to apply to governments. Fiscal conservatives insist that countries should follow the same careful budgetary practices as prudent individuals, but the analogy is questionable because countries differ from individuals in that they can create money, historically by printing it but now by the electronic method that goes by the metaphor of quantitative easing. Modern governments influenced by Keynes allow budget deficits during tough economic times, although they often ignore the recommendation to build surpluses during prosperity, which lead to balanced budgets in the long run. More radically, some modern monetary theorists are enthusiastic about budget deficits as a way of building successful societies.

The fiscally conservative insistence on balanced budgets based on the analogy with individuals employs a weak metaphor because of the crucial difference between the money-producing capacities of individuals and countries. When balancing requires rigid austerity in difficult times, governments lose the capacity to counter the suffering that results from economic disasters such as the Great Depression of the 1930s, the Great Recession of 2008, and the pandemic crisis of 2020. In such cases, the balanced budget metaphor graduates from weak to toxic.

Another dangerous metaphor is balance of payments, which is the difference between money flowing into a country because of financial transactions and money flowing out. Part of this calculation concerns the balance of trade, which is the difference between a country's exports and imports. By analogy to individuals, it might seem that having a positive balance of payments and balance of trade is good for a country, but economists point out that the United States has long had a trade deficit while maintaining substantial economic growth.

Another popular balance metaphor in economics is the recommendation to investors that they maintain a balanced portfolio with appropriate proportions of stocks, bonds, and real estate. Buying and selling to maintain the proportion is called rebalancing. Like advice for individuals to balance their budgets, portfolio balancing is good advice that effectively transfers the positive emotional value of balancing in the body and weight scales. Another sensible metaphor is saying that governments have to set interest rates at levels that balance economic growth against the risk of inflation.

Economic Theory

The most influential balance metaphor in the social sciences is the concept of an equilibrium between supply and demand. In my pleasant neighborhood, houses typically sell for around a million Canadian dollars. Demand comes from people wanting to buy houses like these, and supply comes from the people looking to sell their houses. Prices depend on having buyers willing to pay what sellers are asking, and a sale takes place when the asking amount equals the offering amount.

More generally, figure 8.2 shows how demand can be captured by a line where price declines when many items are available and supply increases when prices are high. At the equilibrium point, the amount supplied matches the amount demanded. This point

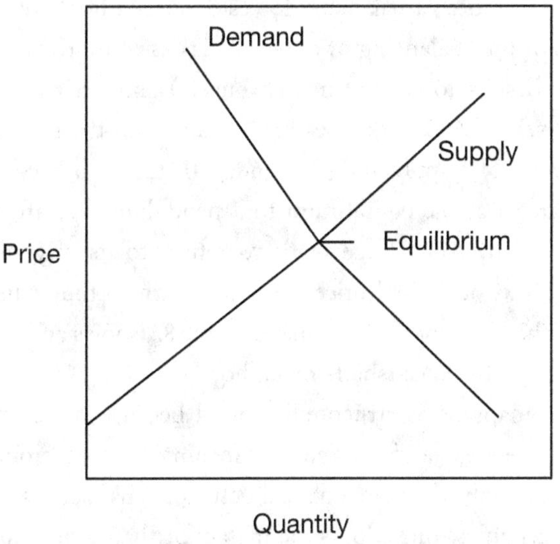

FIGURE 8.2 The demand and supply curves intersect at an equilibrium point that explains the price.

is an equilibrium because of a balance between the two forces of supply and demand that affect price. The value of something for sale does not depend just on what it costs to produce it or on what people want to pay for it, but on the interaction of the two. Such equilibria apply not only to the prices of goods but also to other quantities subject to supply and demand, such as wages and foreign exchange rates.

This pattern of explaining prices originated with Alfred Marshall in his textbook *Principles of Economics.* The book went through eight editions between 1890 and 1920 and became the basis for future textbooks and economic theories. The approach to economics that dominates current universities and business schools consists largely of equilibrium models.

Marshall had a strong background in mathematics and physics, so it is likely that he borrowed the equilibrium idea from the physics idea of balancing forces described in chapter 6. He writes of the "balancing of opposed classes of motives" consisting of desires to acquire goods (buyers) and desires to retain them (sellers).[8] He describes his "inquiry into the balancing of the forces of Supply and Demand."[9] In the eighth edition he compares a stable equilibrium to a pendulum that after being perturbed by some force tends to return to its original position, just as perturbed prices tend to return to the equilibrium established by supply and demand. Table 8.3 shows an analogical mapping behind Marshall's metaphor.

This mapping is structurally sound because the correspondences are clear and consistent. Meaning is more problematic because supply and demand are much more abstract than pushes and pulls unless one adopts the metaphor that desires and fears operate like forces. The equilibrium metaphor accomplished Marshall's purpose; it has provided the basis for 130 years of subsequent investigations, thanks to its elegant explanation of how

TABLE 8.3 Analogy underlying the metaphor of economic equilibrium

Physics equilibrium (source)	Economic equilibrium (target)
Two physical forces such as pushes and pulls	Supply and demand
Movement of object	Prices
Object is stationary when forces are equal	Prices are stable when supply and demand are balanced
When perturbed, the object tends to return to its former position	When perturbed, prices tend to return to their equilibrium point

markets can have stable prices. Most economists would therefore rate equilibrium as the epitome of a strong metaphor.

However, as Marshall recognized, equilibrium explanations assume an idealized world that does not correspond to reality. They apply to situations with a free market of multiple buyers and sellers who make rational decisions. But markets are often dominated by single sellers (e.g., Facebook's monopoly) or single buyers (e.g., Walmart's labor monopsony in some locations). Buyers and sellers have different degrees of access to information about market conditions and therefore will have different expectations about future prices that will affect current prices. Legions of cognitive psychologists and behavioral economists have performed experiments that challenge the assumption that economic decision makers are fundamentally rational.[10]

Most seriously, the emphasis of equilibrium theories on explaining stability limits their ability to explain change, especially the dramatic changes that happen in cycles of booms and busts.[11] Economists struggled to explain the housing and stock market crash of 2008, when prices dropped more than 30 percent in a few months in ways not easily described as changes in supply and demand. Distinguished economists had to resort to poorly specified psychological ideas like "irrational exuberance" and "animal spirits." In my book *Mind-Society*, I show how the strong shifts in attitude that provoke economic bubbles and collapses can be explained by the psychological mechanisms of emotional coherence, including motivated inference in booms and fear-driven inference in busts. Describing prices as falling employs an imbalance metaphor.[12]

The equilibrium metaphor is only dynamic to the limited extent that explains small perturbations in prices rather than large economic changes such as depressions, recessions, and large industrial transformations, which are described by terms like "creative destruction," "disruptive innovation," and "freefall."[13] Economics needs to supplement the balance metaphor of equilibrium with the imbalance metaphor of tipping points. Noticing tipping points such as economic crashes does not explain much by itself because tipping needs to be recognized as resulting from interacting mechanisms, as I argued for climate in chapter 6.

The metaphor of economic equilibrium is embodied to some extent because of its connection to physical forces and weight scales, but it is also transbodied because it uses the abstract concepts of supply and demand, which go beyond the senses. If supply and demand are understood purely in terms of price, then the metaphor is one-dimensional, but additional dimensions are added if the psychological and environmental complexity of supply and demand are taken into account. Economic equilibrium was a novel metaphor in 1890 but has since become conventional through academic adoption.

The notion of economic equilibrium is widely taken to be purely descriptive and neutral, but it has played an important normative role in justifying the status quo. Equilibrium makes stable prices seem like the norm, inheriting the positive emotional values of balance concepts to suggest that free markets are inherently fair. Dramatic economic change that risks the wealth of dominant individuals and companies then counts as instability and as something bad. Equilibrium theories tend to legitimate prices and wages by describing them as the results of the impersonal forces of supply and demand rather than as controlled by agents pursuing economic interests not shared by most of the population. Thus, far from being an emotionally neutral concept of scientific economics, the metaphor of economic equilibrium provides a positive value that supports capitalism. The purpose of the metaphor is purportedly to explain and predict economic events, but it also argues for the appropriateness of a particular form of economy.

My overall evaluation, therefore, is that the equilibrium metaphor in economics is weaker than it seems because it distracts from providing good explanations of important events such as financial crises. Strong metaphors such as natural selection, chemical equilibrium, and cognitive dissonance pave the way for mechanistic theories that explain diverse phenomena, but bad metaphors impede the search for deeper explanations. Moreover, the economic equilibrium metaphor encourages conflation of values and prices, making money the prime measure of goods rather than the extent to which they meet the vital needs of people. The usefulness of economic equilibrium in explaining local price determinations prevents its dismissal as bogus or toxic. But economics could use some new metaphors that are more explanatory and ethical.

Supply-demand equilibrium is the most important balance metaphor in economics, but the concept of equilibrium has been extended to other phenomena. Economists also discuss the occurrence of an equilibrium between intended saving and intended investment. In game theory, a Nash equilibrium is a solution to a game in which the players have nothing to gain by changing their strategies. The players all have to balance their options against what they think are the options of all the other players, with the result that no improvement can be gained. For example, managers of two competing stores are in a Nash equilibrium if neither can figure out how to set prices in a way that makes them more profitable.

Management expert Roger Martin argues that the survival of democratic capitalism requires a better balance between efficiency and resilience.[14] Firms strive for more and more efficiency, losing their ability to adjust to changing contexts such as growing inequality. He fears that without greater resilience capitalism will be supplanted by fascism or communism. As in ecology and personal lives, economic resilience is a balance metaphor because it indicates a return to equilibrium.

Economists also use imbalance metaphors to describe cases where equilibrium is disrupted, using the word "disequilibrium." Severe disruptions are often characterized by falling metaphors such as *collapse* and *freefall.* In 2011, the *Financial Times* described the slide in world financial markets as "economic vertigo."[15]

POLITICS

Politics is the study of government and the state. Like the individuals discussed in the psychology section earlier in this chapter, governments and politicians have to make decisions that balance competing goals and interest groups. Other important balance metaphor in politics are the balance of power among different nations or among different factions within a nation and the operation of checks and balances among branches of government.

Balancing Decisions

Decisions by government leaders and other politicians require trade-offs that are naturally described in terms of balancing. For example, during the 2020 pandemic the *New York Times* described the Korean government in these terms: "The government is also trying to sustain a fragile balance between controlling the virus and safeguarding the economy, and between using government power to protect public health and not infringing on civil liberties."[16] Leaders want to control the virus and also to have a healthy economy, but figuring out how to do both is difficult, as the cartoon in figure 1.1 illustrated. Similarly, governments need to protect public health by mandating actions such as wearing masks and keeping social distance, but democracies naturally worry about putting too much control on their citizens' freedom. People were advised to maintain their balance and remain resilient.

Questions about reopening schools also required difficult trade-offs. The Ontario government wrote: "In planning for the resumption of instruction in the fall, it is critical to balance the risk of direct infection and transmission of COVID-19 in children with the impact of school closures on their physical and mental health."[17] Reopening the schools would be bad if it led to increased spread of COVID-19, not only to the children but also to their teachers and family members. On the other hand, school has many benefits for the children, their families, and for societies in general. No simple formula can capture this trade-off. The balance metaphor works most clearly when two factors are being weighed against each other like the two pans on a weight scale. Putting more weight on one pan pushes up the other and vice versa: you cannot have both pans high or both low.

Balancing health versus education, government actions versus loss of civil liberties, and casualties versus economic growth are all problems without obvious solutions. In lieu of any neat mathematical solution, balancing seems to be the best that we can do. Chapter 10 argues that making these difficult balancing decisions is best if based on a complex and interactive process of constraint satisfaction similar to what is performed by the brain in coordinating the inner ear, eyes, and other bodily inputs. I have collected press reports that use balancing to describe the following conflicts: risk versus reward, safety versus urgency, pandemic versus economics, withholding second doses versus providing more first doses quickly, reserving hospital beds for COVID-19 patients versus maintaining regular services, health versus religious services, empathy versus medicine in care homes, schooling versus disease spread, innate versus adaptive immunity, efficacy versus safety of vaccines, speed versus equity in vaccinations, vaccine risk versus disease risk, caution versus optimism, and physical health versus mental health effects of lockdowns.

However, some countries have developed effective reactions to the COVID-19 epidemic by rejecting the balancing metaphor in favor of extreme efforts to eradicate cases. China, Taiwan, Australia, and New Zealand used strong social controls to largely eliminate the spread of the disease and only then opened the economy, which then could return to normal. In contrast, the balancing approach of countries in North America and Europe has led to a miserable combination of deadly cases and depressed economies. Balancing is not always best and may be toxic if one option really dominates the alternatives.

Another kind of balancing performed by politicians concerns the competing demands of their constituents. Manufacturers, farmers, workers, and other groups push in opposing directions for government actions that will serve their interests. The term "political equilibrium" is used by political scientists in various ways, including the attempts by candidates for office to balance the incompatible demands and values of citizens.[18] Efforts by politicians to deal with competing groups, goals, and interests can be described as a balancing act, striking a balance, or walking a tightrope. If something like the state of the economy is uncertain, we can say that it hangs in the balance. Another way of describing a balance situation uses the economic metaphor of a trade-off: balancing two factors is the same as finding a trade-off between them.

Balance, stability, and equilibrium are politically more valuable than chaos, but some political situations require radical change. If large numbers of the population are coerced and exploited, then revolution may be justified to establish a new regime that can satisfy human rights. In such cases, disequilibrium is desirable.

Other common political balancing acts include freedom versus equality, autonomy versus community, and present versus future. Taxing the wealthy reduces their freedom to use their money as they want but offers the possibility of spreading the wealth more equally in society, which makes more people freer to satisfy their vital needs for food, shelter, and achievement. Because politicians respond to the immediate attitudes of voters, they can be inclined to consider short-term goals such as a successful economy over long-term goals such as a providing a sustainable environment and averting looming disasters resulting from climate change.

Balance of Power

Political balancing usually occurs within one country, but it also has an important international component. Herbert Butterfield summarizes:

> The idea of balance of power initially envisaged the relation between two states . . . as comparable to a pair of scales, with the possibility of intervention by a third party either to restore the equilibrium or to tip the balance in favor of one of the two. Later, the notion was extended, first to three states, then to an entire congeries of states, poised against one another, any substantial change in the mass of one of the units requiring a regrouping amongst the rest if the equilibrium was to be maintained. All this has developed into a wider theory of international politics which makes the preservation of the equilibrium an object . . . of policy for the purpose of preventing the indefinite expansion of a predominant member of the system.[19]

Butterfield uses the metaphors of balance of power and equilibrium to describe the relationships between countries and also to prescribe international behaviors that will maintain the status quo. Tipping the balance is considered dangerous, which shows that the metaphor *balance of power* has an important normative dimension. Emotionally, maintaining the balance of power is good, while losing it and threatening war is bad.

The initial source for the balance of power metaphor was the weight scale with two pans, but the metaphor was generalized to apply to several countries. By the late nineteenth century, leaders were concerned to balance power among five major countries (England, France, Germany, Russia, and Austria-Hungary). Collapse of this balance led to the devastation of the First World War. Table 8.4 shows the analogical mapping between weight scales and the balance of power. Butterfield suggests that the relationships among more than two states could be thought of as involving a balance of forces analogous to the multiple forces that occur in equilibrium in physics and chemistry. He also uses the term "equipoise" as an alternative to "equilibrium."

TABLE 8.4 Analogy underlying the metaphor of balance of power

Weight scale (source)	**Balance of power (target)**
Two pans	Two nations
Weight on each pan	Power of each nation
Balanced when weight is equal—equilibrium	Balanced when power is equal—equilibrium
Adding weight on one side tips scale	Adding power on one side tips balance of power

What is the nature of the power that is balanced in this metaphor? My book *Mind-Society* identifies four kinds of power based on coercion, benefits, respect, and norms. Coercive power uses threats to control the behavior of others through fear of consequences. The second type of power relies on potential benefits such as financial rewards. Respect can also provide power when the leaders of one nation react to the leaders of another nation with liking, admiration, and trust. Finally, norm power operates when one group acquiesces to the plans and goals of another through agreements that seem to them voluntary because they derive from social norms. All four of these kinds of power can operate in the interactions among the leaders of different countries, so the nature of international equilibrium is more like the physical and chemical balancing of different forces than the one-dimensional balancing of weight in a pair of scales.

Maintaining the balance of power requires that the leaders of each nation reflect about how leaders of all other nations are thinking. They need to be aware of one another's emotions, such as the fear in coercive power, the desires in benefit power, the admiration in respect power, and the guilt and pride in norm power. This complexity provides further reason to think of the balance of power metaphor as conflicting forces rather than a single dimension used in a weight scale.

After the Second World War, the balance of power reverted to a two-state conflict between the Soviet Union and the United States. The development of nuclear weapons that could completely wipe out either state introduced mutually assured destruction, which provided motivation to maintain balance. This huge threat gave rise to a new balance metaphor, the balance of terror.[20] The Canadian diplomat Lester Pearson proclaimed in 1955 that the balance of terror had succeeded the balance of power. In the twenty-first century, the rise of China and the European Community shows that balance of power is once again a matter of interactions among many nations and not just two.

Whether the balance of power metaphor is strong or weak depends on the particular historical circumstances. It describes well the situation among the great powers before the First World War and the conflict between the United States and the Soviet Union after the Second World War. With the collapse of communism in 1989, the balance of power metaphor became weaker because the United States turned into the dominant world force, tipping the balance so far that the metaphor became useless for purposes of explanation and prediction. But the economic and military rise of China in the twenty-first century suggests that balance of power is useful again.

Within a country, power is divided among several branches: the legislature, the executive, and the judiciary. In the British parliamentary system these powers are intertwined, but some countries such as the United States emphasize the importance of separation of powers, where each branch has the power to limit the other two. For example, in the United States the president can limit the power of Congress by vetoing its bills, and Congress can limit the president through impeachment. These limitations are characterized by the metaphor of checks and balances, which maintains a balance of power among the legislative, executive, and judicial branches.

Balance metaphors are thus central in politics, including balanced decisions, political equilibrium, balance of power, and checks and balances. Imbalance metaphors are also historically useful, as in describing the decline and fall of the Roman Empire and the downfall of Adolf Hitler. Political balance metaphors have embodied aspects because of their conceptual connections to weight scales and physical forces, but they also are transbodied in their use of abstract, nonsensory concepts such as power, health, and equality. The ubiquity of national and international conflicts and the necessity of dealing with them ensure that political balance metaphors will continue to thrive.

SOCIOLOGY AND ANTHROPOLOGY

Attempts to introduce metaphors of equilibrium and balance into sociology and anthropology have been much less successful than applications to psychology, economics, and politics. The American sociologist Talcott Parsons claimed that all social systems tend to approximate an equilibrium that they will return to when perturbed.[21] But sociologists are more inclined to emphasize social change than social stability.

For explaining change, the imbalance metaphor of tipping points is more useful, especially with changes that are sudden and large. The tipping point metaphor was introduced into sociology by Morton Grodzins in an article on racial segregation in U.S. cities that appeared in 1957.[22] He described a process of Blacks moving from the South and Whites moving to the suburbs, starting with neighborhoods near the center of the city. Following the practice of real estate agents, he called the transition from White to Black occupancy a process of *tipping*, where the *tip point* is the critical level beyond which Whites will no longer stay among Black neighbors. Real estate operators even talk about tipping a building or tipping a neighborhood. An area has racial balance if it has proportional numbers of Whites and Blacks.

The source for the tipping point metaphor might be the human balance system, in which a person can fall over after losing balance, but another likely source is the weight scale, as shown in table 8.5. Tipping the neighborhood from White to segregated Black is analogous to putting so much weight on one side of the scale that it tips over. Sociologists have applied the term "tipping point" to other social phenomena such as attitude change and regime transitions. Malcolm Gladwell popularized the term by describing tipping points in diverse areas such as shoe sales, rumor spreading, crime levels, and teen suicide.[23]

The emotional value of a tipping point depends on whether the change is desirable: a large tip toward lower crime is good, while a tip toward suicide is bad. Tipping metaphors are useful for situations with slow changes with large consequences, as in Hemingway's character in *The Sun Also Rises* who went bankrupt gradually and then suddenly. The tipping point metaphor

TABLE 8.5 Analogy underlying the metaphor of racial tipping points

Weight scale (source)	**Racial segregation (target)**
Two pans	Numbers of Blacks and Whites
Adding weight to one pan	Increasing numbers of Blacks
Adding weight on one side tips the scale	Adding Blacks tips the neighborhood towards all-Black

provides a strong alternative to assumptions of stability and to the popular assumption that large facts have large causes. As with the weight scale, where weights are added slowly, small changes may not seem to have much effect until the system tips over.

Anthropologists debate whether cultures and ecosystems are better described by equilibrium or disequilibrium models.[24] The applicability of balance and imbalance metaphors to cultural groups varies across historical contexts, but some changes seem well described as tipping points. In his book *Collapse*, Jared Diamond describes societies such as the Greenland Norse that seemed to be getting along well until the buildup of external environmental problems and internal political problems led to the rapid extinction of the culture.[25]

LAW, JOURNALISM, SPORTS, FOOD, AND WINE

Other social phenomena are described using balance metaphors, whose use is evident in the law, journalism, sports, and the enjoyment of food and wine. In all these fields, balance is a vivid way of portraying the interaction of opposing factors.

Law

Legal practices are essential to the operations of modern societies, and the law is rich with balance metaphors. The weight scale is the standard symbol used to indicate that the legal system is supposed to be fair because it balances different interests equally. The law is often represented by Lady Justice, a statue of a blindfolded woman holding a balance scale, shown in figure 8.3. Lady Justice

FIGURE 8.3 Lady Justice with weight scales.

Source: Wikimedia Commons.

is derived from the Roman goddess Justitia and the Greek goddesses Themis and Dike, who are also depicted as holding a scale.

The pervasiveness of the weight scale in legal symbolism might suggest that the metaphor is one-dimensional, but in practice legal balance is multidimensional because of multiple interests. In a lawsuit, the interests include those of the plaintiff and defendant, but also of the legal system and other parties such as children in a divorce dispute. Then a balanced judgment in a legal decision is not just a matter of equal weight but also of taking into account multiple interests. These participants can be thought of as conflicting forces or as multiple constraints to be satisfied, analogous to constraint satisfaction in physiological balance, cognitive dissonance, and political decisions.

Smaller balance metaphors contribute to legal discussions. Judges and juries are supposed to consider the weight of evidence and the balance of probabilities in reaching decisions about guilt or innocence. In criminal trials in the British legal tradition that operates in other countries such as Canada and the United States, the accused are considered innocent until proven guilty beyond a reasonable doubt. The term "reasonable doubt" is never defined precisely, but it involves finding a balance between the goal of holding guilty people responsible for their crimes and the goal of avoiding convicting the innocent. Swindlers who have an unfair advantage in a deal can be accused of having their thumb on the scale, a way of shifting the balance in the issue that is immoral and possibly illegal.

Legal balance metaphors are partially embodied through the scale images, but they also go beyond the body with abstractions such as justice and equality. Legal balance metaphors are dynamic because justice requires an ongoing process of evidence presentation and judicial deliberation. As with the body, balance is a positive value associated with legitimate desires for justice and equality, represented conventionally by the weight scale. In contrast to the explanatory purposes of scientific balance metaphors, the main use of legal balance metaphors is persuading people to pursue the appropriate social goal of justice. Failures of legal balance occur because some forces are illegitimate, as when a rich defendant bribes a judge, which is analogous to failure of body balance when the brain gets a bad signal from the inner ear. Legal balance metaphors persuasively encourage good social practices and therefore qualify as strong.

Journalism

A standard prescription of journalism is that reporting should be balanced, but what does this mean? In questions that have just two possible answers, we can think of them as corresponding to the two pans in the scale, with the weight of evidence supporting one answer over another. It then seems appropriate for journalists to report the evidence for each side so that readers can weigh them against each other. A common practice in print journalism, television, and radio is to have reporters talk to advocates of each side, leaving it to the reader to adjudicate. Then journalistic balance in opposition to bias seems as attractive as legal balance.

However, what seems like an admirably fair approach breaks down when the evidence is not close to being equally balanced.[26] For example, when discussing the issue of human-caused climate change, some journalists think that the scientists reporting the evidence in favor of climate change should be counterbalanced by another apparent authority who thinks that climate change is either nonexistent or caused by random fluctuations. The problem is that the vast majority of scientists, more than 99 percent, are confident that the evidence shows that climate change is real and has resulted from human activity through the production of greenhouse gases. Then finding a rare opponent to this position is misleading to the audience because it suggests that the issue is evenly balanced. Similarly, journalists who think they are being balanced in presenting both sides of the debate about whether vaccinations cause autism are ignoring the dominance of scientific evidence against this claim.

Journalists who routinely seek out people to speak on both sides of an issue are rightly accused of suffering from "false balance" or "bothsidesism." Another problem with balanced journalism is that it sometimes focuses on two sides represented by major political parties rather than representing a broader range of alternatives.

In cases such as climate change and debates about whether vaccines cause autism, the balance metaphor for reporting is toxic because it suggests an ongoing debate when the scientific investigations have come down decisively on one side. Journalism faces a tension between balanced reporting and reliable reporting because its goals include finding out the truth as well as being fair to competing sides. So journalists have to find a balance between being accurate and balancing conflicting interests, a kind of metabalance discussed in chapter 10.

Like legal balance metaphors, journalistic ones have value when they persuade people to evaluate arguments based on all the available evidence collected from a variety of authorities. They only become toxic when they are applied blindly in cases such as climate change where the evidence is overwhelmingly on one side. But this problem compels the assessment of journalistic balance as a weak metaphor.

Sports

In sports, the concept of balance is usually literal rather than metaphorical, in line with how chapter 2 describes coordination of the inner ear, eyes, body, and various brain areas. Physiological balance is crucial for sporting activities such as gymnastics (as in the balance beam), riding a bicycle, skiing, and snowboarding. Good balance also contributes substantially to enterprises such as shooting a basketball, hitting a baseball, passing a hockey puck, swinging a golf club, and playing tennis. My favorite shoe company, New Balance, began in 1906 by producing arch supports intended to provide greater comfort and balance.

In addition, balance metaphors contribute to colorful descriptions of sporting activities. A well-balanced team is one that is not dependent on a particular superstar but rather on the interactions of a number of good players. In sports such as football and basketball, successful teams have a good balance between offense and defense, which means that they are equally good at scoring points and at preventing the opposite team from scoring points. Teams that start losing all their games are described as suffering a collapse, an imbalance metaphor that suggests a dramatic fall. Sports leagues aim for a competitive balance where various teams contend for championships. Hence balance is both a literal and metaphorical component of athletic activities.

Food and Wine

Chapter 7 mentioned the important medical metaphor of a balanced diet, and a balanced meal is a specific combination of the different foods needed to promote health. Beyond health, food provides pleasure through a metaphorical balance of complementary ingredients.[27]

One of my favorite cuisines is Thai food, which pays special attention to questions of balance. Thai dishes succeed by integrating tastes that are sweet, sour, spicy, salty, and bitter. Sweetness can come from fruits and sugar, sourness from lime and tamarind, spiciness from chili peppers and peppercorns, saltiness from fish sauce, and bitterness from leaves. Too much of any one of these ingredients makes a dish overwhelming, as too sweet or too sour, but balance promotes pleasure. With multiple ingredients, balanced food is too complicated to map onto the one-dimensional weight scale, so it corresponds better to richer kinds of balance such as interacting forces or the multidimensional sensory inputs that produce physiological balance.

Similarly, a balanced wine combines five key components: alcohol, acidity, sweetness, tannin, and fruitiness. Wines with too little of these components will seem bland, but ones with too much of a single component such as acidity will be hard to drink.

The best wines, such as a well-made Cabernet Sauvignon, are greater than the sum of their parts, producing a Gestalt experience where the whole combination of the five components can make the drinker ecstatic. Similarly, a balanced beer combines the sweetness of malt and the bitterness of hops to generate flavor, aroma, and mouth feel to produce an overall positive experience. The most likely source analog for these beverage balances is the physics one of multiple balancing forces.

Sophisticated diners strive for a balance between balances when they look for food and beverage pairings that bring out the best in all the ingredients. With Thai food, I prefer to drink beer, although a Riesling wine with a touch of sweetness can also be pleasant. Experts propose such principles as balancing acidity in wine with fat and sweet foods and combining sweet wine with salty foods. Red wines have more tannins and flavors than white wines and so tend to go better with hearty food. Wines are often characterized by weight metaphors such as "heavy" and "light," and the general recommendation is to pair heavy wines with heavy food and light wines with light food. The advice is complicated by the fact that some red wines such as Pinot Noir are lighter than some white wines such as oaked Chardonnay.

Balance is always emotionally positive—no one praises a dish or drink for its overwhelming dominance by one factor. Eating and drinking are intensely sensuous, so balance metaphors for food and wine are fully embodied. Their use may be conventional for experts but novel for newcomers to the intricacies of fine dining and imbibing. The primary purpose of talking of food and wine in terms of balance is to encourage good combinations of components that bring people pleasure, and the metaphor provides good advice and also explains why some combinations work. Accordingly, I mark food and wine metaphors as strong.

BALANCING SOCIETY

Balance metaphors pervade society, from everyday psychology to law and journalism. Scientific investigations in psychology, economics, politics, and sociology also employ balance metaphors with varying degrees of success. Metaphors that have contributed most to social explanation include cognitive balance and dissonance in psychology, the balance of power in politics, and the scale of justice in the law. Flawed balance metaphors with undue influence include economic equilibrium and balanced reporting. For sociology, the imbalance metaphor of tipping points provides an emphasis on change rather than stasis. Good social theorists look for tipping points that mark substantial changes such as economic collapses and political revolutions.

Society as a whole is a balancing act among the competing interests held by individuals, groups, and countries. Balancing metaphors are often useful ways of describing the decisions that must be made every day by ordinary people and by political leaders. Normatively, we need to watch for cases where balance metaphors serve merely to reinforce the status quo rather than bring about substantial change. For example, balanced reporting is an attractive goal until it gets in the way of finding social truths that can help people to live their lives in ways that promote their needs through deeper kinds of sensemaking.

My examination of social balance has neglected creative practices in the arts. Let us now look at the importance of balance in literature, film, music, and painting.

9

THE ARTS

Balance operates physically in artistic activities such as dance and drama that require movement. But in most arts, balance operates figuratively through metaphors that adorn literature, film, music, and painting. I explore the power of balance metaphors by discussing novels by Rohinton Mistry and Jean-Paul Sartre, Hitchcock's film *Vertigo*, musical harmony, and paintings by Leonardo da Vinci and Raphael. These works use balance metaphors that range from one-dimensional comparisons based on weight scales to comparisons with multiple constraints analogous to balancing the body. The creative interaction of balance and imbalance metaphors enriches aesthetic experience.

LITERATURE

Many authors use balance metaphors in their fiction and nonfiction writings. A particularly rich employment is by Rohinton Mistry. Mistry was born in Bombay and immigrated to Canada in 1975 when he was twenty-three years old. His second work, the prize-winning and best-selling novel *A Fine Balance*, was published in 1995.

The novel has four main characters (Dina, Maneck, Ishvar, and Omprakash) who deal in India with problems of poverty, housing, politics, and social relationships with families and friends. The title comes from a statement made to Maneck by a proofreader he meets on a train: "You see, you cannot draw lines and compartments, and refuse to budge beyond them. Sometimes you have to use your failures as stepping-stones to success. You have to maintain a fine balance between hope and despair. . . . In the end, it's all a question of balance."[1] Much later in the book, the proofreader has become a lawyer and offers another character similar advice: "There is always hope—hope enough to balance our despair. Or we would be lost."[2]

Each of the characters encounters potentially overwhelming causes of despair, including bereavement, financial insecurity, physical threats, abysmal living conditions, homelessness, and dismemberment. Nevertheless, they trudge through the book with a degree of hope that their lives will get better. The most upbeat moment in the book occurs when the four main characters are living together in a crowded apartment that nevertheless offers them a degree of security and social support.

The metaphor of a balance between hope and despair concerns emotional balance, which I mentioned in chapter 8 in discussing everyday psychology. Both hope and despair are attitudes toward the future, with hope an emotionally positive attitude that things will get better and despair the emotionally negative opposite that things will continue to be bad or even get worse. For the proofreader/lawyer, the existence of a balance between hope and despair is an emotionally positive metaphor because it implies that even when you are suffering despair, a chance exists that events will tip you back to the emotional state of hope.

Because hope and despair are the only two elements balanced in Mistry's metaphor, its source is most plausibly the weight scale, as shown in table 9.1. Just as a weight scale can tip back and forth depending on the weights put on the two pans, so a person's emotional state can tip back and forth depending on the occurrence of good and bad events. This metaphor is dynamic rather than static and deals with situations that can produce swings between different emotional states. It is one-dimensional, with the positive/negative range of emotion corresponding to the weight on the scale's pans. Embodiment is evident in the metaphor because of the forces operating on the weight scale and the physiological correlates of emotions such as hope and despair, but the abstract concern of these emotions with the future reveals transbodiment as well.

What is the purpose of Mistry's balance metaphor, which he mentions three times in the book, including in the title? The main characters do not explicitly express belief in a balance between hope and despair, but their actions in continuing with their often-disastrous lives suggest that they believe it implicitly. Despite enormous threats to their autonomy, well-being, and physical integrity, none of them gives up until the end of the book. Instead, they keep on striving with sometimes desperate

TABLE 9.1 Analogy underlying the metaphor of balance between hope and despair

Weight scale (source)	**Hope and despair (target)**
Two pans	Two emotions, hope and despair
Weight on each pan	Amount of hope or despair
Adding weight on one side tips the scale	Events tip minds between hope and despair

strategies to persevere. I take Mistry's lesson to be that even when life seems overwhelmingly bad, it has possibilities of improvement that justify some degree of hope, although the unhappy endings of his characters suggest a note of irony in his title.

My first reaction to the metaphor of balance between hope and despair was puzzlement because of the oddness of balancing emotions. But I now think that Mistry's metaphor is superb because he uses it to describe a profound emotional experience that also happens to people whose lives are not as difficult as those of the characters in his novel. For example, the pandemic of 2020–2022 made many people struggle with illness, social isolation, work insecurity, and even the threat of homelessness. But it helped to realize that the pandemic would not last forever and that many people's health, relationships, jobs, and housing situations would eventually improve. Despair tends to feed upon itself, with sad thoughts provoking more sad thoughts. Mistry's balance metaphor can help people to remember that they have had hopeful times in their lives that can reasonably be expected to recur.

Other pairs of emotions can metaphorically be interpreted as balanced: happiness and sadness, confidence and surprise, pride and shame, gratitude and resentment, forgiveness and anger, enjoyment and disgust, calmness and fear, and even love and hate. According to the theory of emotions mentioned in chapter 4, emotions are brain representations that combine a situation, physical changes in the body as a response to the situation, and appraisals of the relevance of the situation to a person's goals. From this perspective, forming an emotion is a constraint satisfaction process analogous to the account of balance in chapter 2.

For example, if you see a speeding car coming toward you, your body will react with physical changes such as rapid heartbeat while your brain evaluates the situation as a possible threat to your existence. The result is intense fear that lasts until you recognize that the car has missed you, your body state returns toward normal, and you reevaluate the situation as lacking in threat. Like all neural states involving constraint satisfaction, tipping points can naturally move you from one state to another, in this case from fear to calmness. Describing the emotional configuration as a balance and emotional changes as tipping points are metaphors that capture important aspects of how emotions work in the brains and lives of people in the world. Therefore, I think that Mistry's metaphor of a balance between hope and despair qualifies as strong, not just for those emotions but for emotions in general.

The book has a few other balance metaphors concerning a balanced diet, equilibrium as a return from a state of unease, and an oppressed minor character believing that the scales will never balance fairly. Another minor character decides that it is wiser to despair rather than have stupidly raised hopes, challenging the metaphor of a balance between despair and hope.

An additional metaphor repeated in Mistry's book has nothing to do with balance. Dina uses the scraps from her sewing business to make a quilt that becomes a metaphor for life. Maneck says that God is a quilt maker who is working on a giant cloak with a pattern that became so big and confusing that he abandoned it. The story concisely conveys the feeling of despair that God is no longer paying attention to the lives of suffering people. Later, a lawyer's complex tale is compared to an endless chain of silk scarves pulled from the mouth of a conjurer. Dina had intended the patchwork quilt as a wedding present, but it ends up as a dirty and fraying cushion on the wheelchair of a legless beggar. Mistry does not say so explicitly, but the balance is clearly tipped from hope to despair. Maneck concludes that God is a bloody fool with no notion of fair and unfair who could not read a simple balance sheet, and he jumps in front of a train.

Then balance becomes a metaphor for justice, in line with the scales of justice discussed in chapter 8.

Another novel with a balance metaphor in its title is *Nausea* by Jean-Paul Sartre.[3] Published in 1938, it was the first expression of the philosophical view of existentialism, which became influential in the 1950s. The main character, Roquentin, has frequent occurrences of a nausea that is clearly more than an upset stomach, as it also has vertigo symptoms such as spinning colors. In chapter 3, I described balance-related nausea as an inclination to vomit resulting from a mismatch among sensory signals from the inner ear, eyes, and body. Similarly, Roquentin becomes nauseous because of a mismatch between his dismal reality and cultural expectations of a meaningful life based on work and love: he does historical writing that he has come to find pointless and suffers from failed romantic relationships. He also longs for adventure as a contrast to his dismal provincial life and feels that his existence is absurd, dominated by the emotions of disgust, anxiety, exhaustion, and despair. Roquentin finally resolves his distress not by killing himself like Mistry's Maneck but by deciding to move to Paris and write a nonhistory book. His freedom then becomes exciting rather than oppressive.

Table 9.2 maps Sartre's nausea metaphor as a comparison between Roquentin's psychological distress and physiological nausea. Sartre's metaphor is richer than Mistry's because it compares two kinds of conscious experiences (weight scales are not conscious) across multiple dimensions of sensory inputs and expectations about life. Mistry's scale source is conventional, but Sartre's nausea source was novel. Both metaphors are dynamic because they describe changes in people's lives, and both are intensely emotional. Sartre blends embodied experiences of visceral nausea with transbodied philosophical concerns about the meaning of life.

TABLE 9.2 Analogy underlying Sartre's metaphor of nausea

Physiological nausea (source)	Psychological nausea (target)
Mismatch among sensory inputs	Mismatch between expectations and reality
Feelings of imbalance and stomach distress	Feelings of disgust, anxiety, and despair
Mismatch causes unpleasant feelings	Mismatch causes unpleasant feelings

For me, Sartre's nausea metaphor is not as compelling as Mistry's balance between hope and despair. Existentialism was an overreaction to the demise of a religious worldview that insisted that meaningful lives depend on divine prescriptions. The secular world can be meaningful rather than absurd through the balanced pursuit of the natural human goals of love, work, and play, as I argue in chapter 10.

FILM

Alfred Hitchcock's *Vertigo* is often ranked as one of the best films of all time, but what does it have to do with vertigo? The lead character Scottie, played by James Stewart, suffers from dizziness connected with fear of heights, but he also seems afflicted by two metaphorical kinds of vertigo connected with social relationships and incoherence. Scottie's vertigo has much in common with Roquentin's nausea.

Hitchcock's 1958 movie was based on a 1954 French novel called *D'entre les morts* (From among the dead). It begins with the detective Scottie chasing a criminal across the roof in San Francisco and almost falling to his death. This trauma gives him acrophobia, fear of heights, which manifests as severe dizziness described as vertigo. After retiring from the police force, Scottie is asked by an old friend Gavin to follow his wife, Madeleine, who is obsessed with her deceased great-grandmother. Madeleine is a striking blonde played by Kim Novak, and Scottie falls in love with her after saving her life when she jumps into the sea by the Golden Gate Bridge.

The first half of the movie is a slow and ridiculous depiction of Scottie's infatuation with the increasingly deranged Madeleine. When she climbs a tower in a mission outside of San Francisco, he is unable to follow her because of his fear of heights. After she falls to her apparent death, he becomes so disconsolate that he has to spend months in a mental hospital.

After Scottie recovers, he remains obsessed with Madeleine and keeps seeing women who look like her. He approaches a look-alike redhead named Judy Barton who works in a shop. It turns out that Judy, also played by Kim Novak, is in fact the woman who impersonated Madeleine as part of a plot by Gavin to get away with murdering his wife, who is actually the person thrown from the mission tower. But Judy as Madeleine has fallen in love with Scottie and so she agrees to see him. In his obsession, Scottie insists that Judy change her clothes and hairstyle to duplicate Madeleine's.

Hitchcock uses cinematic techniques to convey Scottie's severe balance problem. The opening credits show a swirling eye that foreshadows his vertigo. When Scottie looks down from the roof in the first scene, a novel camera shot that later became known as the "Vertigo effect" shows his discomfort by having the camera zoom in on the ground while moving back from it. The same shot is used when Scottie fails to climb the tower after Madeleine. The emotional distress of vertigo is also accentuated by pulsing strings of the intense accompanying music. Scottie is also shown as having balance problems when climbing a stepstool in the apartment of his friend Midge.

From the perspective of the scientific account of vertigo presented in chapter 3, Scottie's diagnosis of vertigo is bogus. He does not seem to suffer from the spinning feeling that is the hallmark of vertigo as opposed to general dizziness. Hitchcock does suggest that spinning is a problem with his swirling eye in the credits and with some special effects used before Scottie's hospitalization, but otherwise Scottie's problem seems to be generic dizziness. Moreover, his so-called vertigo is given a Freudian-style explanation as having resulted from the trauma of falling from the roof at the beginning of the film. The official explanation is that the fall caused his acrophobia, which causes vertigo, but it would be more plausible to think that the vertigo causes acrophobia. Unlike the known causes of vertigo described in chapter 3, I know of no cases where a single traumatic event has caused vertigo. Finally, the psychiatrist in the mental hospital where Scottie is placed makes the nonsensical claim that Scottie's vertigo can only be cured by another traumatic event, which does happen later in the movie.

Nevertheless, Hitchcock clearly liked the theme of vertigo and chose the title of the film over alternatives preferred by the studio. The philosopher Robert Pippin suggests in a book about the film that Hitchcock used vertigo as a metaphor for Scottie's severe problems with interpreting other people.[4] That Scottie has this difficulty is evident from his relationship with Midge, to whom he was engaged in college; no explanation is provided for why she broke off the engagement but nevertheless remained friendly with him. Scottie's relationship vertigo gets even worse when he has to deal with Madeleine, who seems bizarrely occupied with her great-grandmother but oddly interested in him.

More specific disruption happens to Scottie when he sees Judy put on a necklace that had been worn by Madeleine, a replica of a necklace in a portrait of her great-grandmother. Suddenly, Scottie realizes that Judy is not just a facsimile of Madeleine that he has created, but is in fact Madeleine! This generates an astonishment that metaphorically amounts to vertigo. The trauma is so great that, in line with the earlier Freudian prediction, Scottie overcomes his vertigo and is able to follow Judy up the mission tower when they return to it.

Pippin suggests that Scottie has a problem reconciling competing needs for dependence and independence, a common relationship issue of emotional incoherence. The conflict between vital human needs for autonomy and relatedness generates a failure of constraint satisfaction analogous to the conflicts among sensory inputs that generate vertigo and nausea.

Hitchcock largely flubbed the science of vertigo, but he powerfully portrayed the balance disorder of dizziness triggered by heights. Just as effectively, he presented metaphorical vertigo provoked by uncertainty in romantic relationships and especially by astonishing events that have no explanation. Scottie's shocks included seeing Madeleine fall off the mission tower not just once but twice and finding out that Judy was really Madeleine. Surprise is the emotional reaction that arises when events do not fit together coherently.

Scottie's head spins both from fear of heights and from being overwhelmed by the astonishing events that occur. Watching the movie generates similar astonishment when the last twenty minutes reveal that Judy was pretending to be Madeleine and was complicit with Madeleine's husband in his wife's death. Although Hitchcock's vertigo theme has little to do with the balance disorder of vertigo, it nevertheless works as a strong

TABLE 9.3 Analogy underlying Hitchcock's metaphor of vertigo

Physical vertigo (source)	Relationship vertigo (target)
Mismatch among sensory inputs	Mismatch between romantic expectations and reality
Feelings of imbalance and spinning	Feelings of confusion, surprise, and distress
Mismatch causes unpleasant feelings	Mismatch causes unpleasant feelings

metaphor for the sometimes-overwhelming uncertainty of romantic relationships, as shown in table 9.3. Like Roquentin's nausea, Scottie's vertigo results from a dynamic, conscious mismatch between what he thinks and what he learns.

The various kinds of vertigo in Hitchcock's movie are beautifully portrayed in the official poster shown in figure 9.1. The swirling loop captures the vertigo theme, while the two unbalanced bodies display the dizzying uncertainty of the film's relationships.

Another film with a balance theme is the 2002 science fiction movie *Equilibrium*, which takes place in a future world where war has been ended by giving everyone a drug that eliminates emotion. The term "equilibrium" is the name of the building where people are given the drug to make them emotion-free. People who refuse the drug in order to have emotional experiences are guilty of "sense offence" and severely punished.

The use of equilibrium in this movie is highly misleading. As chapter 8 describes, emotional equilibrium is not the absence of emotions but rather having emotions balanced in appropriate proportions. Mistry in his novel was not looking for the

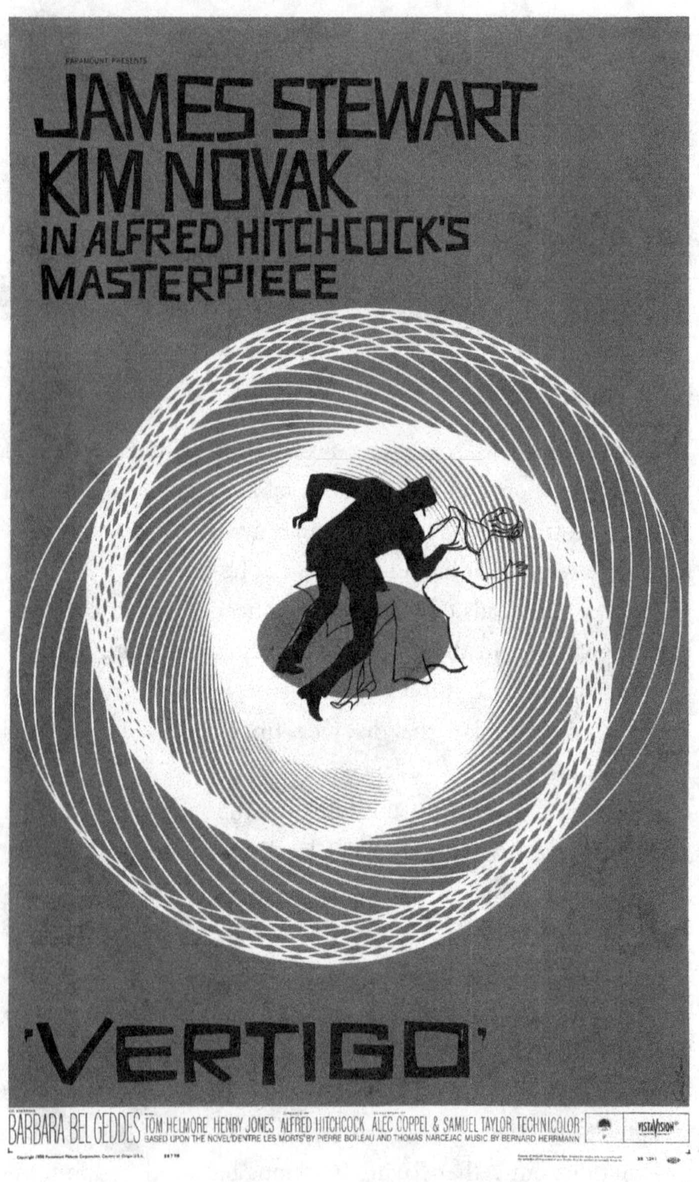

FIGURE 9.1 Official poster for the movie *Vertigo*.

Source: Wikimedia Commons.

obliteration of hope and despair but rather having them operate in tolerable proportions. The movie would be better described as oblivion rather than equilibrium, so the balance metaphor fails, in contrast to the power of Hitchcock's vertigo metaphor.

The film (and musical play) *Fiddler on the Roof* is organized around a metaphor that spans balance and imbalance. Its characters keep their balance by maintaining tradition, without which their lives would be as shaky as a fiddler on a roof.

MUSIC

The psychology discussion in chapter 7 pointed out connections between balance metaphors and musical ones. Balancing attitudes requires replacing cognitive dissonance with consonance, where both balancing and dissonance are based on emotional coherence: the processing of multiple conflicting constraints introduces initial discomfort that is happily resolved.[5] Some musical balance metaphors connected with harmony have similar emotional complexity, but others are much simpler.

The simplest balance metaphors in music use a single dimension that needs to be regulated. With stereo speakers, the volume needs to be balanced so that the same amount of sound comes from both the left and the right. Volume also requires balancing to ensure that both a singer and accompanying instrument are audible. Sound from a boom box is balanced through appropriate levels of bass and treble. In these cases, a weight scale furnishes the metaphorical source, and adjusting the volume or bass/treble is analogous to adjusting the weights on the two pans. These simple metaphors maintain the convention that balance is good and imbalance is bad. The purpose of making the adjustment is to improve the auditory experience of the listener.

These metaphors are slightly dynamic in that repeated adjustments may be required to get the balance right when the goal is a stable state with good sound.

More complicated metaphors operate with multiple performers and instruments.[6] In his MasterClass video, Itzhak Perlman says, "I'm a big worrier about balance," by which he means that he does not want his violin to be overwhelmed by an orchestra. A string quartet sounds best with a balance in volume, pitch, and timber among the two violinists, the viola, and the cello. Barbershop quartets have four singers, including a lead, tenor, baritone, and bass singing in four-part harmony. More generally, any band or orchestra will sound better if the contributors are balanced in both volume and tones, such as lightness and heaviness. Such balance can be adjusted by the musicians' playing and by their placement in the room. Auditory engineers use equalizing techniques to change the balance of different frequency components in a recorded signal.

Unlike the one-dimensional balance scale, the metaphorical source for the balance metaphors that operate with multiple performers is multidimensional, producing a constraint satisfaction problem. The positive constraints are that the voices and instruments should be sufficiently similar in tone, volume, and timing to fit well with one another. The negative constraints are that they need to be sufficiently different that they can contribute different components to the overall sound. Like the beautiful paintings discussed next, great music requires unity in variety.

The most sophisticated balance metaphors in music concern harmony, which is the combination of complementary sounds that please the listener. Most simply, harmony combines notes into a chord, for example, the notes C, E, and G into the C major chord that is popular in Western music. These notes are consonant in that they sound pleasant to many people, for physiological reasons concerning how ears and brains work, and for cultural reasons resulting from people's experiences and expectations. Dissonant chords such as the combination of C and C-sharp sound harsh and unpleasant. The playing of dissonant chords sets up a tension (metaphorically, a kind of psychological stress) that listeners expect to be resolved by a return to enjoyable consonant sounds. In another balance metaphor, consonance is described as stable, while dissonance is unstable.

A pleasantly harmonized piece of music has a balance between consonant and dissonant sounds. Dissonance surprises and challenges the listener and keeps the music from being repetitive and boring. For example, Beethoven's *Ninth Symphony* opens with a shocking combination of a B-flat inserted into a D minor chord. Balance is restored by a return to melodies in the keys of B-flat major and D major, which are associated with heroic triumphs.

Some twentieth-century composers such as Arnold Schoenberg tried to tip the balance toward dissonance with atonal music, but this achieved far less popularity than traditional classical music. Alternative music such as post-punk, which experimented with replacing harmonies by atonalities, had tiny success compared to the harmonies of groups such as the Beatles, who blended the voices of John Lennon, Paul McCartney, and George Harrison.

The description of harmonization as balancing consonance and dissonance is a metaphor that is much more dynamically and emotionally complex than the other balance metaphors in music. The metaphor is far from static because it concerns the succession of dissonances and consonances that can unfold over a whole piece. Compare the static balance of two people, each getting half a sandwich, with the dynamic balance of people sharing chores by taking turns doing the dishes. As in the other music metaphors, the balance between consonance and dissonance is emotionally positive, but it incorporates mixed emotionality by including and resolving the disturbing unpleasantness of dissonance.

My book *Natural Philosophy* remarks how "melodies can gain coherence by means of familiar rhythms, arrangements of ascending and descending notes, compatible timbres (sound quality of notes), and varied repetitions."[7] A higher kind of coherence occurs in songs where the melody and lyrics are well matched with each other, as in the Beatles' song "Yesterday," which combines a melancholic melody with wistful words. Putting them all together produces emotional coherence through simultaneous satisfaction of multiple constraints. Table 9.4 shows the general structure of the metaphor of musical balance. I cannot claim that most people employ this rich mapping, but it sums up the simpler mappings that people use when talking about balance in music.

TABLE 9.4 Analogy underlying metaphors of musical balance

Bodily balance (source)	**Musical balance (target)**
Sensory inputs from inner ears, eyes, and body	Inputs of sounds from instruments, players, and speakers
The brain makes sense of these inputs by satisfying multiple constraints	Musicians make sense of combinations of sounds by satisfying multiple constraints
Balance results from successful constraint satisfaction and is unconscious	Balance results from successful constraint satisfaction and is pleasurable
Imbalance results from failed constraint satisfaction and is unpleasant	Imbalance results from failed constraint satisfaction and is unpleasant

PAINTING

Painting similarly depends on balance by constraint satisfaction. According to Shelley Esaak: "Artists generally strive to create artwork that is balanced. A balanced work, in which the visual weight is distributed evenly across the composition, seems stable, makes the viewer feel comfortable, and is pleasing to the eye. A work that is unbalanced appears unstable, creates tension, and makes the viewer uneasy."[8] Leonardo da Vinci's *The Last Supper*, shown in figure 9.2, achieves balance by using symmetry in both the background and the portrayal of the apostles. Unlike the two sides of the background, the two groups of apostles on either side of Jesus are not identical, but six similar apostles on each side maintain balance.

Mark Johnson provides an illuminating discussion of the importance of symmetry in painting and other visual arts.[9] He points out that different manifestations of balance are based on a symmetrical arrangement of forces. For example, if we carry

FIGURE 9.2 Leonardo da Vinci's balanced painting, *The Last Supper*.

Source: Wikimedia Commons.

an equal load in each of our hands, then the force on our left hand is the same as the force on the right hand. Walking also involves symmetric forces because, if we feel ourselves falling in one direction, we can exert force with our legs or move our arms to produce a symmetrical force in the other direction. Even homeostasis within our bodily organs has symmetry because, if we feel discomfort from too much gas in our stomachs, we can apply an opposite force to cause a belch to relieve the discomfort.

Johnson draws on the work of Rudolf Arnheim to suggest that all visual perception has a structure of tensions and forces.[10] A picture with one disc placed in the center of a square is inherently balanced, but if the disc is moved to one side or if another disk is added, then people have to see the picture as balanced or unbalanced. The desired balancing is psychological rather than physical, in that an image does not have physical weights but rather provokes the desire for a distribution of visual weights. The factors that influence visual weight and balance in the work include location, spatial depth, size, intrinsic interest, isolation, shape, knowledge, and color. All of these contribute to how the different parts of the painting appear symmetrical and hence balanced by having roughly equal forces on each side.

The visual importance of symmetric forces does not mean that painters want to make both sides of a work identical. Just as music gains from a combination of dissonance and consonance, painting gains from a balance that overcomes differences. My favorite theory of beauty is Francis Hutcheson's idea that works are beautiful if they have uniformity in variety, combining internal diversity to make them interesting with enough coherence among the different parts so that the work hangs together overall.[11] Symmetric visual weight in a painting where the two sides are not the same provides such unity and variety, allowing balance to be a major contributor to beauty.

FIGURE 9.3 Raphael's balanced painting, *The School of Athens*.

Source: Wikimedia Commons.

Figure 9.3 shows one of my beloved paintings, Raphael's *School of Athens*. Plato and Aristotle center the scene with other great thinkers arranged around them. Unlike the building in the background, the symmetry of the people is not perfect because of different individuals and groups, but it is sufficient to provide a sense of unity. The philosophers in front of Plato and Aristotle are Heraclitus (writing at a desk) and Diogenes (lounging on the steps). The group on the left centered around Pythagoras (writing in a book) is balanced by the group on the right centered around Euclid (writing on a slate).

Balance in paintings as approximate symmetry is too rich to be captured by a one-dimensional source such as the weight scale. Rather, the source for pictorial balance is multidimensional like physical balance, depending on combinations of sensory inputs. As Johnson and Arnheim indicate, balancing requires taking into account location, spatial depth, size, intrinsic interest, isolation, shape, knowledge, and color, all of which furnish constraints that are approximately satisfied to furnish a sense of perceptual equilibrium. The two sides of a balanced painting are analogous but not identical.

Sculptures can similarly involve symmetric forces but operating in three dimensions, as in the reclining figures of Henry Moore. Symmetry is even more important in architecture because most buildings are well balanced both visually and physically so that they do not tip either in reality or in our perceptions. For example, Gothic cathedrals and modernist skyscrapers adhere to strict standards of symmetry and balance. A major exception is the work of architect Frank Gehry, whose constructions such as the Guggenheim Museum in Bilboa use highly unusual materials such as titanium formed into flowing shapes that violate the expectations of Western architecture. In his MasterClass video, Gehry proclaims that "imbalance is nice." Like Schoenberg and modern jazz musicians, Gehry surprises and interests observers by violating the usual expectations of symmetry and balance.

One limitation of the Johnson and Arnheim account of symmetry and balance in art is that they neglect its emotional component. We have frequently seen that balance produces an emotionally pleasing effect, whereas imbalance can be unpleasant or even disturbing. As I suggested for harmonious music, the pleasure of balance results from emotional coherence where everything fits together. Similarly, the vast majority of paintings, sculptures, photographs, and buildings in the Western tradition display large amounts of balance that contribute to emotional coherence.

But familiarity, predictability, and pleasantness are not the only characteristics that make art interesting. My book *Natural Philosophy* gives examples of paintings that are disturbingly ugly, such as works by Picasso, de Kooning, Francis Bacon, and Goya.[12] The central imperative of all kinds of art is not just to be beautiful but to also be emotionally engaging, and art that is shocking captures people's attention and sometimes gains their esteem. For example, many of Picasso's cubist paintings violate the symmetry of human faces with results that are far from beautiful but nevertheless fascinating. So even though balance is generally emotionally good and imbalance is generally bad, artists sometimes employ violations of symmetry to make their products more eye-catching.

BALANCING THE ARTS

I have exhibited the importance of balance metaphors in four kinds of artistic endeavor: literature, film, music, and painting. Some metaphors build on simple, one-dimensional sources, as when Mistry's balance between hope and despair aligns with the weight scale. More commonly, balancing is better identified with the multidimensional constraint satisfaction that operates in balancing the whole body while standing or walking. Sartre's *Nausea* and Hitchcock's *Vertigo* are built on complex mappings between their character's distressed psychological states and physiological imbalance. Harmony in music and balance in painting are also multidimensional.

In artistic domains, balance is usually emotionally positive and imbalance is negative, but exceptions show how the unusualness and disturbance of imbalance add notes of surprise and interest to a work of art. Beethoven's symphonies, Hitchcock's *Vertigo*, and Gehry's architecture incorporate balance and imbalance, a process that the next chapter discusses as metabalance. Artistic sensemaking combines balance and imbalance metaphors to enhance creativity and aesthetic enjoyment.

10

PHILOSOPHY

Philosophers are not renowned for their balance, except for the Philosophy Department in Tom Stoppard's hilarious play *Jumpers*. These philosophers are selected for their acrobatic ability because the vice chancellor of the university is an indifferent philosopher but an excellent gymnast.

The verbal gymnastics of philosophers have sometimes employed balance metaphors, most notably the idea of reflective equilibrium, which has been influential in ethics. The study of morality has employed other balance metaphors, such as Aristotle's golden mean. Normative discussions of the meaning of life stress the importance of balancing conflicting purposes such as love and work. Balance metaphors have also contributed to the nature of knowledge, as in finding a balance between the goals of acquiring truths and avoiding falsehoods.

Philosophy is usually classed with the humanities rather than the sciences, but many philosophers since Aristotle and Hume have seen close connections between philosophy and science. I describe such overlaps in my book *Natural Philosophy*, which nevertheless distinguishes philosophy from science as being more general and more normative. Each science is concerned with its particular domain, for example, biology with living things. But the concerns of philosophy have the highest generality, such as the question of what kinds of things exist, ranging from atoms to gods. In addition, science is largely concerned with descriptive questions about how things are, although applied sciences such as engineering occasionally deal with normative questions about how things ought to be. Philosophy, in contrast, is always intensively normative in its investigations of how people ought to act and how knowledge can best be gained.

Figure 10.1 provides a rough placement of various fields on the two dimensions of generality and normativity. Physics has high generality since it concerns all physical entities from atoms

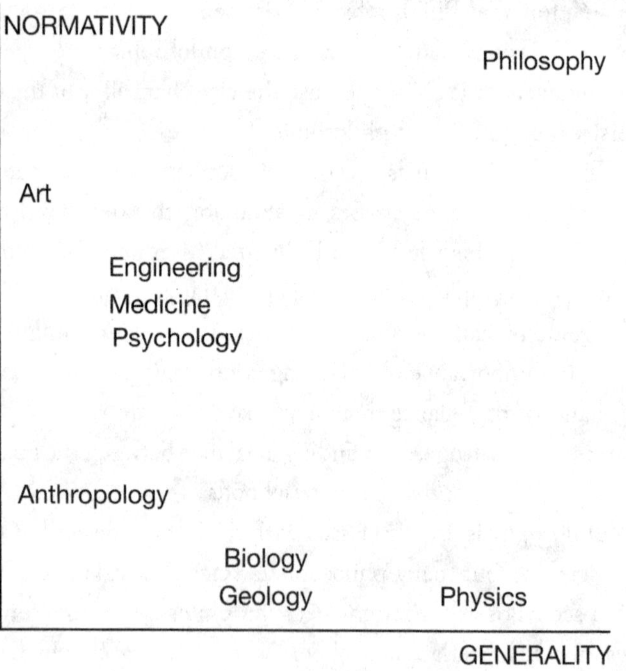

FIGURE 10.1 Location of fields on the two dimensions of generality and normativity.

to galaxies. But it has low normativity because it has no interest on how atoms and galaxies ought to be. Geology and biology more specifically concern structures and life forms on particular planets, paying little attention to moral judgments. Anthropology is even less general because it concerns human cultures on just one planet, and it has only a small normative component because anthropologists generally try to avoid moralizing about the cultures they study.

Engineering, medicine, and psychology all depend on descriptive studies of the operations of structures, bodies, and brains in this world. But they also have a substantial normative component because they aim at building better structures such as bridges, healthier bodies, and minds that benefit from education and psychiatric treatments. I have placed art as having minimal generality because it is largely concerned with human activities, but I have identified it as high in normativity because of ambitions to produce beautiful paintings and sublime music.

Philosophy rates high on generality and normativity but has to deal with tensions between them when general descriptions clash with normative aims. Posing impossible norms is futile and pointless, as when demanding that people always act perfectly or believe the infinite logical consequences of their beliefs. Normativity should take into account the insurmountable limitations of people's minds and bodies, while at the same time pushing people to exceed the mediocre and despicable behaviors that people often display, such as using metaphors that are bogus or toxic. Metaphorically, philosophy accomplishes balancing acts between the descriptive and the prescriptive, between generality and normativity. In complicated cases in ethics and epistemology, the best way to do this is to convert balance metaphors into standards of constraint satisfaction as a more rigorous approach to sensemaking.

KNOWLEDGE

In epistemology (the philosophy of knowledge), the dominant metaphor is that knowledge has foundations like a building whose lower portions are grounded in the earth and provide the basis for structures built upon them.[1] Philosophers have disagreed about whether the foundations consist of sense experience (empiricists) or of self-evident truths achieved by reason alone (rationalists). Trenchant critiques have shown that neither foundation serves its intended purpose, and nonfoundational alternative metaphors have been suggested: knowledge is more like a cable of interwoven strands, a ship on the ocean that is rebuilt as it sails, or a crossword puzzle with interconnecting clues.

When the foundation metaphor is abandoned, we can notice trade-offs in the development of knowledge in science and ordinary life that can be described by balance metaphors. The aims of knowledge include believing truths and rejecting falsehoods, but pursuing these aims requires balancing the risks of making mistakes against the benefits of learning useful truths.[2] A sure way to avoid falsehoods is to believe only the most obvious truths, but they provide little of the information needed to conduct one's life and to make difficult decisions about health, relationships, and work.

Consider medical decisions, such as when to approve vaccines for treatment of the scourge of COVID-19. Applying the highest standard of absolute truth concerning efficacy and harmlessness would mean that no vaccine would ever be approved and people would remain helpless against the pandemic. But applying a low standard could easily lead to faulty adoption of useless or harmful vaccines. As with the legal issues described in chapter 8, authorities must balance evidence for and against hypotheses concerning the effectiveness of particular vaccines. David Hume recognized this problem when he described what to do in the face of conflicting evidence: "We must balance the opposite experiments, where they are opposite, and deduct the smaller number from the greater, in order to know the exact force of the superior evidence."[3]

The goals of knowledge are not just achieving truths and avoiding falsehoods, since many truths, such as the number of stones in my driveway, are too trivial to pursue. We want to acquire truths that are important to us because they help us to satisfy our goals, such as explaining what is going on and fulfilling vital needs such as food. Scientists do not collect data about random phenomena but rather about happenings that are relevant to their theoretical and practical interests. Hence achieving knowledge requires not just balancing evidence but also balancing interests to decide what data and theories are worth pursuing. The goals of knowledge include explanation, prediction, and practical applications as well as gaining truths and avoiding falsehoods, all of which have to be balanced against one another.

Particular examples of theory choice are also balancing acts because they need to take into account alternative theories, the full range of evidence, and background theories that provide mechanisms that explain why the theories might be true. For example, the hypothesis that global warming is increasing because of human emission of greenhouse gases has to compete with the hypothesis that warming is just normal temperature fluctuation that is independent of human action. But human-caused warming is much more plausible than the random fluctuation hypothesis because it explains many more facts, such as that warming has steadily increased with greenhouse gas emissions. Moreover, a good mechanism explains why these emissions lead to global warming: through the trapping of heat by the atmosphere. Attending to this conclusion is strongly in humanity's interest because huge suffering will result if warming passes the tipping point and leads to irreversible sea-level rises and extreme weather events. Unfortunately, some business and government leaders follow their narrow interests and deny humans' controllable role in global warming. Making good political decisions about global warming requires balancing theory and evidence along with sound values.

For characterizing knowledge, balance is a much stronger metaphor than foundations, but it would be stronger if it could be translated into a mechanism that specifies precisely how to handle trade-offs among competing theories and conflicting evidence. For one-dimensional balancing such as the weight scale, the mechanism is simple—just apply more weight or other quantity to the two sides until they are even. Multidimensional balance problems are much harder to solve because they require weighing different factors simultaneously.

Chapter 1 introduced an analogy between tackling balance problems and solving picture puzzles such as the one in figure 10.2. Solutions require fitting hundreds of pieces together to produce the full picture. Even matching two pieces together is tricky because they could fit in five ways (left, right, top, bottom, or not at all) and there are four criteria for matching them (color, shape, continuity of objects represented, and formation of edges). For a 1,000-piece puzzle, the number of possible ways of putting pieces together exceeds the 10^{80} particles estimated to occur in the whole universe. Nevertheless, people routinely solve such puzzles in less time than a soccer star is reported to have done when he was proud about doing one quickly even though the box said seven to eight years.

Figuring out how all the pieces join together is a constraint satisfaction problem. Positive constraints are ones you want to satisfy, as when accepting both that one piece fits with another

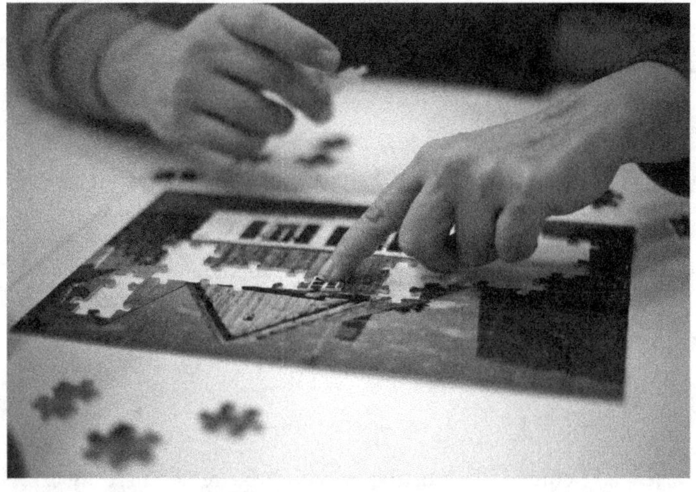

FIGURE 10.2 Picture puzzle.
Source: Tobias Klenze/Wikipedia Commons.

and that they have matches in color and shape. Negative constraints are ones that you do not want to satisfy, as when you cannot accept both that one piece fits to the left of another and that another piece fits in the same place: a puzzle piece can only dock with one other piece on a given side. Figure 10.3 shows a constraint satisfaction network with positive constraints (solid lines) between fits and matches and negative constraints (dotted lines) between competing fits.

In general, constraint satisfaction problems are computationally intractable, which means that as the problem size gets larger the solutions become exponentially harder to find. Fortunately, efficient algorithms can produce good approximations. The most psychologically natural algorithms use neural networks to perform constraint satisfaction in parallel. For a neural network implementation, the nodes in the network are represented by

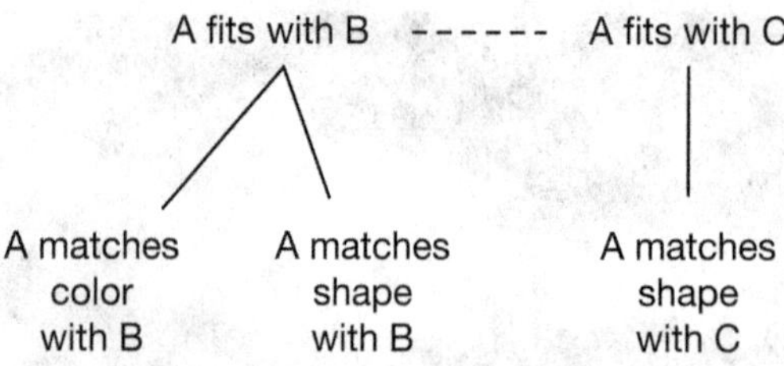

FIGURE 10.3 Constraint satisfaction network concerning whether a puzzle piece A fits with B or with C. The guess that A fits with B explains the matches in color and shape (*straight lines*), but the guess conflicts (*dotted line*) with the guess that instead A fits with C.

artificial neurons, the positive constraints are captured by excitatory links, and the negative constraints are captured by inhibitory links. Spreading activation around the neurons by means of the excitatory and inhibitory links activates some neurons and de-activates others. This result is interpreted as accepting some nodes and rejecting others, thereby solving the constraint satisfaction problem.

In figure 10.3, think of each guess as being represented by a single neuron, such as one for that A fits with B. The plausibility of the guess is represented by the firing rate of the neuron—the faster the neuron firing, the greater the plausibility of what it represents. The effect of excitatory links shown by solid lines is that if one neuron fires, then the other neuron also fires. The dotted lines are mutually inhibitory in that each neuron's firing tends to suppress the firing of the other neuron. Simple mathematical rules adjust the firing rates of each neuron based on the firing rates of each neuron to which it is connected by excitatory or inhibitory links. Repeated cycles of such adjustments usually lead to stable firing rates in which some guesses are accepted (their neurons are firing fast) and some are rejected (neurons firing slowly).

The puzzle analogy helps us to understand balance in the complex multidimensional case. A puzzle solver has to solve constraints about how to fit two pieces together and how to fit pieces together. Similarly, to achieve, maintain, and regain balance, the brain has to coordinate information coming from parts of both inner ears; the body, including arms and legs; the two eyes; and multiple brain areas, including the brainstem, the cerebellum, and various parts of the cortex. Balancing is solving a huge puzzle with many pieces. Both balancing and puzzle solving are examples of constraint satisfaction problems that can naturally be managed by neural networks.

In particular, arriving at knowledge requires balancing theories, evidence, and alternative theories to reach a coherent result. This process might seem mysterious, but balancing, puzzle solving, and other mental processes can be understood as satisfying multiple constraints, a process effectively accomplished by neural networks. Since 1989, I have published articles and books that apply a theory of explanatory coherence based on constraint satisfaction to many cases of scientific, medical, and legal reasoning, from scientific revolutions to COVID-19.[4] The end of this chapter describes a general method for converting balance metaphors into constraint satisfaction problems.

RATIONAL DECISION-MAKING

Richard Thaler replied to the question of how he planned to spend the million dollars from his 2017 Nobel Prize in economics by saying: "I will try to spend it as irrationally as possible."[5]

Thaler's prize was awarded for decades of research showing that people are much less rational than economists have assumed.

Chapter 8 presented the role of balance in decision-making from a purely descriptive perspective, but philosophy is concerned with the normative question of how people ought to make decisions. According to economists, people should make decisions by using probabilities and expected values to calculate what choices maximize expected utility, but in real-life situations people rarely know the relevant probabilities and utilities. Important decisions about love and work have to made with much cruder estimates of goals and consequences relevant to determining whether to get married, get divorced, have a baby, pursue a career, change jobs, or retire. To be made well, such decisions require several kinds of balancing.

All complex decisions require tensions and trade-offs between incompatible goals. Getting married potentially helps to accomplish goals such as having company, affection, sex, emotional support, and a family. But marriage can also threaten other goals, such as autonomy, work success, and serenity. No neat mathematical calculation can reconcile these conflicts. The balance metaphor is a useful way of indicating that the different factors have to be weighed against one another without any crisp algorithm.

Like acquiring knowledge, making hard decisions is analogous to solving a picture puzzle when putting the pieces together requires satisfying positive constraints about color, shape, objects, and edges while respecting negative constraints about not fitting two pieces together in the same way. Similarly, you cannot both get married and not married, and the decision requires dealing with positive constraints such as belonging versus freedom. Solving constraint satisfaction problems requires feeding the options and constraints into a coherence engine such as the human brain.[6]

Emotions are an important part of this process because the strength of goals and the intensity of constraints are marked not by simple numbers but by emotional reactions that combine neural perceptions of bodily changes with appraisals of whether a goal is relevant. For example, the combination of intensive physiological reactions and cognitive judgments about a partner may add up to intense love that swamps negative considerations about marrying. Usually, however, important decisions concern multiple goals, so that the balance metaphor is multidimensional and cannot be reduced to weighing one quantity.

Personal decision-making can be irrational in various ways, most commonly by neglect of important goals. A person who feels that being married is the ultimate goal or that work success is all that matters is blocked from making decisions that lead to the fully meaningful lives discussed below. A trenchant *New Yorker* cartoon shows an old man on his deathbed proclaiming, "I should have bought more crap!," an ironic complement to the observation that no one dies wishing to have spent more time at the office. Rational decision-making benefits from thinking of life goals as factors to be balanced rather than as absolute requirements.

This kind of balancing in personal decisions concerns one person dealing with multiple simultaneous goals, but complications arise when decision-making concerns goals at different timescales, goals of different people, and trade-offs between risk and caution. People are often irrational when their decisions require balancing immediate and long-term goals. Here are some examples from the mundane to the globally serious.

- I want to eat this rhubarb pie right now, but I don't want to gain weight.
- I want to buy this jacket, but I should be saving for old age.

- I want to heat my home comfortably, but I worry that global warming is rapidly increasing because of greenhouse gases.
- I want to party with my friends, but I know that such gatherings increase the spread of COVID-19.

These cases require balancing immediate desires against long-term goals.

In such cases, people often form intentions to act in their long-term interests but succumb to impulsive desires, like Oscar Wilde, who said that he could resist anything except temptation. Philosophers describe cases where desire conquers reason as weakness of will, while psychologists call them intention-action gaps. People may trick themselves into believing that their short-term goals are actually more important than their long-term ones, a process known to philosophers as self-deception and to psychologists as motivated inference.

Philosophers and economists who follow Aristotle in assuming that people are basically rational have trouble understanding why people commonly succumb to weakness of will and self-deception. Neuroscience does much better because it appreciates that decisions and other inferences depend on interactions among brain areas that are inaccessible to consciousness. Short-term desires such as wanting pie result from the firing of neurons in pleasure-oriented brain areas such as the nucleus accumbens. In contrast, long-term planning with abstract goals such as wanting to be healthy results from the firing of neurons in the prefrontal cortex.[7] So the problem of balancing desire and reason can be reframed as a problem of balancing the nucleus accumbens and the prefrontal cortex. Such balancing occurs through parallel constraint satisfaction carried out by networks of neurons that excite and inhibit one another, an adjustment process far more complex than simply adding up weights.

Another problem in rational decision-making is how to balance your own interests against the interests of other people. According to the philosophical doctrine of egoism espoused by Ayn Rand, rationality requires exclusive attention to one's own goals. But psychology recognizes such exclusive attention by diagnostic terms such as "narcissist," "psychopath," "sociopath," and "antisocial personality disorder." Fortunately, such people are rare, only around 1 percent of the population. At the other extreme, the philosophical doctrine of utilitarianism insists that ethics requires appreciating the pleasure and pain of all people as much as we appreciate the pleasure and pain of ourselves, our family members, and our closest friends. Moral saints who can value the interests of people they have never met as highly as those of themselves and their family members are probably even rarer than psychopaths.

Beyond these extremes, we can be genuinely altruistic in caring about other people while recognizing that it is impossible for most of us to care as much about strangers as we do for ourselves and those close to us. The task then is to find a balance between the interests of self and others, taking all of them as constraints on rational decision-making. This balance metaphor qualifies as strong because it helps us to understand the complexity of making decisions when people have conflicting interests.

Difficult decisions also require balancing risk and certainty. Psychological experiments find that people have a bias toward choices that come with high certainty, but rationality sometimes requires choosing something risky to get a high payoff over a safer choice that has a lower payoff. For example, over the long run, investing in bonds has a high probability of a low payoff, while investing in stocks has a lower probability of a high return on investment. Going for certainty may reduce anxiety but also reduce long-term pleasure from higher gains.

Ecological decisions are sometimes limited by a precautionary principle that warns against introducing any technology that has a chance of being dangerous. But Freeman Dyson has responded: "The Precautionary Principle says that if some course of action carries even a remote chance of irreparable damage to the ecology, then you shouldn't do it, no matter how great the possible advantages of the action may be. You are not allowed to balance costs against benefits when deciding what to do."[8] Worrying about the risks of ecological threats such as global warming is justified, but balancing risks and benefits is always a better strategy than avoiding risks altogether. If early humans had followed the precautionary principle, they would still be hanging out in caves in central Africa.

Thus rational decision-making requires at least four kinds of balancing: among personal goals, between immediate and long-term goals, between one's own goals and those of others, and between certainty and risk. In all of these cases, balance metaphors are useful for steering people away from simplistic methods of decision-making such as the economic principle of maximizing expected utility.

Chapter 2 introduced constraint satisfaction by using an example of decision-making about eating a steak or a vegetarian burger. In general, decisions can be interpreted as constraint satisfaction problems because they represent actions and goals by elements that are connected by positive and negative constraints.[9] An action that accomplishes a goal establishes a positive constraint between the action and the goal, as when getting a high-paid job accomplishes a goal of having money. Negative constraints are between incompatible actions, for example, taking a high-paying job over another that is more interesting. As with constraint satisfaction problems about knowledge, decision problems can easily be translated into neural networks that solve them at least approximately.

MORALITY

Moral choices are even more normative than rational decisions because they concern whether the choices conform to ethical standards of right and wrong. Ethical decisions are complicated because different standards can support different choices; this problem requires striking a balance between conflicting principles, rights, and virtues. For example, the ten commandments of Judaism and Christianity tell people both to honor their parents and not to steal, but they do not tell people what to do if the parents are starving and the only way of feeding them is to steal food.

Moral Balance

Many ethical traditions have proposed that balance is useful in acting morally. Aristotle did not explicitly use balance metaphors in his doctrine of the golden mean, but his advocacy of finding the desirable middle between extremes is often construed as a kind of balancing.[10] For example, courage taken to excess amounts to recklessness, so people need to find a balance or mean between too much and too little courage. Similar principles have been advocated by Confucius, Judaism, and Hinduism. The Hindu idea of karma suggests that the universe has a balance that uses reincarnation to reward or punish people for current behavior: what goes around, comes around. Given the lack of evidence for reincarnation and karmic adjustment, this kind of balance is bogus.

Some contemporary philosophers and psychologists have similarly advocated moral balance as the method that people use for determining right and wrong. According to Mordecai Nisan, people calculate a moral balance for themselves based on all their relevant actions within a time period and compare this balance to a personal standard.[11]

Balancing is also required for applying moral principles, as with the prominent approach to medical ethics that operates with four principles:[12]

1. Autonomy: respect people's freedom.
2. Beneficence: provide benefits to people.
3. Nonmaleficence: avoid harm to people.
4. Justice: distribute benefits, risks, and costs fairly.

Application of these principles when they conflict requires difficult balancing. For example, people who refuse to get vaccinated against COVID-19 are exercising autonomy but failing nonmaleficence because they might spread the disease.

Similarly, Kantian ethics based on duties often has to deal with conflicting duties, as when I was once asked by a first date whether she looked like the picture she had posted on her dating site: I had to balance the duty to be truthful against the duty to be kind. Virtue ethics tells people to act in accord with good character traits such as being honest and being kind and encounters similar balance problems when different aspects of character suggest different results.

My country Canada has joined a few other countries in allowing medically assisted suicide for people wanting to end ceaseless suffering. Ethical and legal debates have concerned how to frame the law to strike a balance between aiding people to escape horrible lives and protecting people with disabilities who might be vulnerable to coercion.

These moral balance metaphors are valuable in showing the complexity of ethical decision-making, in contrast to monolithic recommendations to do what God wants or to calculate what action provides the greatest pleasure for the greatest number of people. But they only qualify as weak because they provide no effective means for balancing between conflicting principles, duties, or actions. The metaphorical dizziness that comes from failures to find ethical balance has been called moral vertigo.

Reflective Equilibrium

Where do ethical principles come from? Religion says they are delivered by God as revealed in holy books such as the Bible and the Quran. A more secular but still absolute approach says that moral rules are self-evident, so we can acquire them by reason alone. Disappointingly, neither religion nor pure reason has delivered ethical principles that have earned general agreement.

John Rawls, the most influential moral and political philosopher of the last hundred years, proposed an alternative method that he called *reflective equilibrium.* This idea is the latest in the chain of equilibrium metaphors that I have surveyed, from physics to chemistry to biology to economics. Rawls's method consists of finding a balance between general moral principles and particular judgments about specific cases. For example, a general principle might be that a just society should have a fair distribution of wealth, which fits with a particular judgment that the United States is unjust because some people have much more wealth than others. Reflective equilibrium is a dynamic process in which principles and judgments are adjusted until they balance: "A conception of justice cannot be deduced from self-evident premises or conditions on principles; instead, its justification is a matter of the mutual support of many considerations, of everything fitting together into one coherent view."[13]

TABLE 10.1 Analogy underlying the metaphor of reflective equilibrium

Economic equilibrium (source)	Reflective equilibrium (target)
Supply and demand	Principles and particular judgments
Prices	Agreements on principles and judgments
When prices are unstable, adjust supply and/or demand until they stabilize	When disagreement occurs, adjust principles and/or judgments to make them fit
Eventually, stable prices result	Eventually, agreement results

Rawls's writings display an extensive understanding of economics, so it is plausible that the source analog for this metaphor is the idea of economic equilibrium discussed in chapter 8. Table 10.1 shows the mapping between the equilibrium of supply and demand and the equilibrium of principles and judgments. Prices can fluctuate because of variations in supply and demand, but increases and decreases in supply and demand will eventually lead to stable prices. Similarly, repeated adjustments in principles and judgments can eventually lead to accepted moral rules. This balance metaphor is multidimensional because many different principles and judgments need to be adjusted together.

The balance metaphor shown in table 10.1 is superior to the rigid religious and pure-reason approaches to choosing ethical principles. The comparison is dynamic because it may require many iterations of principles and judgments before equilibrium is reached, and in later work Rawls acknowledged that equilibrium may be elusive. He also came to recognize that equilibrium is not just among moral principles and judgments but can also require integration with empirical facts from fields such as psychology and economics. This more extensive integration is called *wide* reflective equilibrium, and its incorporation of empirical information makes it even more multidimensional than the narrow kind restricted to principles and judgments.

Despite its influence, reflective equilibrium does not qualify as a strong metaphor because it has substantial problems.[14] The theory of justice as fairness that Rawls arrived at has received broad respect but only narrow acceptance. Identification of reflective equilibrium as a method for reaching a stable state has not decreased the intense disputation in ethics and political philosophy. We saw a similar problem in chapter 8 with economic equilibrium: that it fails to explain the instability of prices evident in crises such as depressions and wild inflation. Although wide reflective equilibrium provides a bit of an anchor in the world by incorporating empirical information, this anchor is not sufficient to bring agreement. In contrast, the natural sciences achieve much consensus through systematic reliance on experiments and observations.

Another problem with reflective equilibrium is that it can arrive at stability too easily. People who are dogmatic or under the sway of authoritarian personalities or groupthink may reach equilibrium in their principles and judgements despite their deficiencies. For example, Donald Trump seems to have no doubts about his moral judgments, principles, and empirical assumptions, which are highly coherent with one another. The failure of reflective equilibrium to produce reasonable conclusions is analogous to economic equilibrium when prices stabilize because monopolistic firms or authoritarian governments dominate the economy. The particular judgments that go into the interactive process of arriving at moral principles may be useless intuitions based merely on social prejudices rather than anything indicative of moral truth. Experimental philosophy has found that people's intuitive judgments vary with gender and cultural background.

Moreover, the reflective equilibrium metaphor has failed to flesh out the vague idea of coherence between principles and judgments. Coherence has to be more than consistency, but a stronger explanation of how things fit together needs to be given. The theory of coherence as constraint satisfaction and its neural network algorithms would be a great help, but its application to moral principles and judgments has never been worked out.

The expanded method of wide reflective equilibrium promises to gain some objectivity by integrating empirical findings into moral considerations, but no one has specified how this actually works. In contrast, I have argued that psychology and biology provide important findings relevant to morality that concern vital human needs.[15] Such needs are different from transient wants in that they concern the biological and psychological factors required to function fully as a person. Most obvious are biological needs such as oxygen, water, food, shelter, and health care. But substantial evidence supports the existence of psychological needs of relatedness to other people, competence to achieve valued goals, and autonomy from interference.[16] These needs are commonly satisfied by the pursuit of love, work, and play. Vital needs anchor moral principles much more effectively than the mere back-and-forth process of reflective equilibrium. The procedure of defending moral rules should be based on needs and coherence construed as parallel constraint satisfaction, not the underspecified achievement of equilibrium. Hence the balance metaphor of reflective equilibrium is bogus.

Following Nelson Goodman, Rawls defended reflective equilibrium in ethics as analogous to the procedures used in justifying principles of deductive and inductive logic, where principles and inferential practices are incrementally adjusted to one another. But logical principles should be subject to the same standards I applied to knowledge acquisition in general: truth, avoidance of falsity, explanation, and practical importance. These goals serve to evaluate logical principles in the same way human needs serve to evaluate ethical principles. In *Natural Philosophy*, I recommended this normative procedure:[17]

1. Identify a domain of practices, such as knowledge or ethics.
2. Identify candidate norms for these practices, such as logic or utilitarianism.
3. Identify the appropriate goals of the practices in the given domain, such as truth or human welfare.
4. Evaluate the extent to which different practices accomplish the relevant goals.
5. Adopt as domain norms those practices that best accomplish the relevant goals.

Following this procedure should produce better principles of logic and ethics than reflective equilibrium.

THE MEANING OF LIFE

Albert Camus said that the only serious philosophical problem is: why don't you kill yourself?[18] He was exaggerating as philosophy has many serious problems about knowledge, reality, and morality, but he was pointing to the importance of the question of whether life has meaning or purpose. A sense of meaningless is a major part of deep depressions that lead people to consider suicide.

Answers to the question of what makes life worth living can be nihilistic, monistic, or pluralistic. The nihilist answer is that life is inherently meaningless and people would be better off if they had never been born. This desperate conclusion is countered by the observation that people in most countries report their lives as being moderately satisfying. When not sunk in dismal poverty, most people find aspects of their lives that mark them as happy and meaningful.[19]

Most simply, meaning arises from just one factor, such as religion. When I was a child, I learned from the Catholic catechism that God made me to know, love, and serve him in this world. Then meaning accrues purely from a relationship with God. More secular monistic views of the meaning of life can come from exclusive obsession with a career, family, or sports team. Monistic views of the meaning of life make decisions easy because you only have to consider one factor in deciding what direction your life should take.

Most people, however, have multiple sources of meaning in their lives, such as my favorite trio of love, work, and play. Pluralistic meaning is psychologically natural because research identifies the vital needs mentioned earlier for relatedness, autonomy, and competence. Love, work, and play are effective contributors to the satisfaction of these psychological needs. But pluralism introduces a serious balancing problem because of the need to deal with trade-offs between conflicting values and needs.

The metaphor of finding balance in meaningful lives is strong, multidimensional, and dynamic. It can be amplified using the constraint satisfaction account of balance developed in chapter 2. Just as literal, physical balance is a matter of satisfying conflicting constraints provided by sensory and memory signals, the metaphorical balance required for having a meaningful life requires satisfying multiple constraints that derive from vital needs.

These needs are objective and universal in the lives of all humans, although there may be individual and cultural variations. Some cultures such as East and South Asian societies put more emphasis on social connections than on personal freedom, so that they stress relatedness over autonomy. The weighting of needs changes in the different stages of people's lives; for example, play is of greater importance for children and retirees. In contrast, people with young families and growing careers usually focus on love and work.

The objectivity of these psychological needs and values is further supported by identifiable brain mechanisms, for example, the activation of pleasure areas such as the nucleus accumbens by romantic interests.[20] Neural and molecular mechanisms explain why social isolation cripples lives by impacting the amygdala, ventral striatum, and orbitofrontal cortex, including epigenetic changes in gene expressions that can contribute to heart attacks. Your brain and your culture are both important in establishing your life as meaningful in accord with your multiple needs.

The multiplicity of needs makes balancing a good way of thinking about the meaning of life. The metaphor highlights the pluralistic nature of life's meaning and suggests how to deal with unavoidable conflicts in the factors that make for good lives, such as love, work, and play. People need to balance their lives in roughly the same way that they need to balance their moving bodies, by satisfying multiple external and internal constraints. Table 10.2 lays out the profound analogy that underlies the strong metaphor of balance in meaningful lives.

TABLE 10.2 Analogy underlying the metaphor of balance in meaningful lives

Physical balance (source)	Balance in life (target)
Goal: stable standing and walking	Goal: meaningful life
Constraints from sensory inputs and memories	Constraints from needs and values
Coherence through constraint satisfaction leads to effective standing and walking	Coherence through constraint satisfaction leads to well-being and happiness
Incoherence leads to falls, dizziness, and distress	Incoherence leads to distress and depression

METABALANCE

In the 2018 movie *Avengers: Infinity War*, the villain Thanos says about a knife: "Perfectly balanced, as all things should be." In contrast, we have seen many realms in which balance is not always desirable, from nature to architecture to philosophy. My last normative (and therefore philosophical) question concerns metabalance, by which I mean finding balance between balance and imbalance. A web search for "metabalance" turns up references to dodgy nutritional supplements and exercise programs. Instead, I use metabalance to mean balance about balance, specifically the question of how to perform trade-offs between balance and imbalance. Usually, balance is good for people and imbalance is bad, but this book has identified numerous cases where different kinds of imbalance have advantages.

The clearest example of the importance of metabalance is art and architecture, where balance is largely the symmetry that contributes to the beauty and usefulness of paintings and buildings. But perfect symmetry can be boring, and introducing asymmetries generates interest and surprise. For example, the *School of Athens* painting has great symmetry in its general composition but much variation in the depiction of the philosophers on either side of Plato and Aristotle. The architect Frank Gehry has produced startling buildings in line with his doctrine that imbalance is nice, but the underlying core of buildings such as concert halls and museums inevitably requires much internal symmetry to be functional. The characterization of beauty as uniformity in variety recognizes the value of combining coherence and incoherence. No absolute standard for how to reconcile balance and imbalance is available, but their integration is crucial for producing art and artifacts that satisfy goals such as beauty, emotional engagement, and practical usefulness. Similarly, musical pieces are more stimulating when they balance consonance and dissonance.

Chapter 8 mentioned the economic problem of balancing efficiency and resilience. Resilience is a balance metaphor because it concerns returning a system to equilibrium. Efficiency is a kind of stability, but resilience shows the importance of being able to respond to external surprises by regaining stability. Hospitals are good examples of the tension between efficiency and resilience: high efficiency means that their resources of beds and staff are fully used, but this efficiency can limit their ability to deal with unexpected disasters such as the COVID-19 pandemic. We cannot identify in general the best balance between efficiency and resilience, but it is easy to identify cases where institutions have gotten the balance wrong and cost lives. Journalists have a serious metabalance problem when they try to balance fairness (construed as balancing competing interests) against the obligation to get things right.

Similarly, leading a meaningful life requires finding a balance between balance and imbalance. Systematically pursuing the satisfaction of appropriate needs and values is a good strategy for achieving meaning, but sometimes it is fun to be spontaneous, wild, and crazy. The regular, predictable life can benefit from occasional ventures into irregularity. I do not have a formula for determining how much imbalance is good because people's circumstances vary. Spontaneity is more to be valued in a new romantic relationship than in raising a sick child. It is enough to recognize that balance is not always good and that moderating it with surprising imbalance can make life livelier. Challenging the golden mean, we can strive to be moderate about everything, including moderation. Philosophy should not aim for the comfortable balance of reflective equilibrium when it can pursue norms about thinking and behaving that improve human lives.

One of the great powers of the human mind is the recursive ability to think about thinking. Chapter 5 mentioned the analogy about analogy that compares it to Winston Churchill's description of democracy as the worst system of government except for all the others. Talk of toxic or dead metaphors is itself metaphorical. The recursive capacity of the human mind is limited because we get lost if we have to consider too many layers of aboutness, but it nevertheless enables us to be creative in applying concepts to themselves. Metabalance itself is a strong, novel metaphor because it illuminates previously unrecognized complexity in how we think about our lives.

How can metabalance problems be resolved? They require reconciling constraints about constraints, which stretches the recursive capacity of the mind to its full extent. As with ethics, my recommendation is to regain focus by concentrating on human needs, both the physiological needs such as food and the psychological needs such as love and work. In trying to figure out when to go for balance or imbalance, we should keep in mind the slogan: need, not greed.

BALANCED PHILOSOPHY

Understanding balance in both its literal and metaphorical applications requires a view of thinking different from the commonsense conception based on verbal communication that is serial, proceeding one step at a time. Talking and writing work serially because sentence follows sentence. The verbal-serial mode of thought is enshrined in intellectual accomplishments such as Aristotle's syllogisms, Euclid's geometrical proofs, Bertrand Russell's formal logic, and computer programming languages.

Brains support verbal-serial inference when they generate language, but they are also capable of a more basic mode consisting of multimodal parallel constraint satisfaction, which I have called sensemaking. Sensemaking is multimodal rather than verbal because in addition to words it operates with representations that include sensory inputs based on external senses such as vision and internal senses such as pain, perceptual images such as faces and terrains, and emotions such as happiness. Sensemaking is parallel because the human brain operates with around 86 billion neurons firing simultaneously rather than one at a time. This parallelism allows brains to solve problems by using groups of neurons to represent elements and constraints among them, considering the elements and constraints simultaneously rather than step by step.

Sensemaking is more basic than verbal-serial inference because it occurs in nonhuman animals that lack language, in prelinguistic infants, and in all humans when they engage in tasks that are sensory, perceptual, and emotional. Literal physiological balance and most kinds of metaphorical balance are better construed as sensemaking than as verbal-serial inference.

Chapters 2–4 explained body balance as the result of neural mechanisms that make sense of information from the inner ear, eyes, body, and numerous brain areas. Sensemaking is multimodal in combining diverse representations of signals that concern fluid motion in the semicircular canals, visual perceptions, body location and movement, and memories of previous experiences. The brainstem interacts with other areas such as the cerebellum to interpret these signals by taking them as constraints on how to explain the diverse signals. Neural groups in different brain areas interact in parallel to figure out a good way to satisfy the various constraints and arrive at a coherent interpretation that maintains balance.

Sometimes constraint satisfaction fails and imbalance occurs in the form of dizziness, vertigo, nausea, or falls. Sense-making can be derailed in numerous ways, such as by bad signals from the inner ear and bad processing by damaged brain areas. Restoring good balance works by fixing bad signals: doing the Epley maneuver to get errant calcium crystals out of the inner ear canals, or retraining the brain with exercises such as tai chi that make it better at interpreting bodily inputs. Consciousness is also sensemaking because the brain identifies which of its countless representations in different modalities are sufficiently important to command attention and generate experiences.

Balance metaphors operate with other kinds of multimodal parallel constraint satisfaction. Understanding that living well balances love, work, and play requires grasping the analogy between life events and the problem of satisfying constraints among things that matter. Analogy itself is a constraint satisfaction problem because finding a mapping depends on considering semantic relations between the elements of the source and target, the relational correspondences between the source and target, and the purposes of the analogy and the metaphor it drives. Evaluation of a metaphor as strong, weak, bogus, or toxic requires assessing whether it accomplishes its purpose.

In this chapter, balance metaphors in philosophy range from strong (balancing meaning in life, metabalance) to weak (moral balance) to bogus (reflective equilibrium). I have not found any toxic balance metaphors in philosophy, which does contain toxic analogies, as when Aristotle justified slavery by saying that slaves are like women in being naturally inferior.

In principle, every balance metaphor can be converted into a constraint satisfaction problem by the following Conversion Procedure.

1. Identify the source and target analogs for the metaphor, such as life as the target and balancing the body as the source.
2. Identify the elements in the target that need to be balanced, such as the activities of love, work, and play.
3. Identify the positive constraints that affect the elements, for example, that love requires deep relationships with other people.
4. Identify the negative constraints that affect the elements, for example, that love and work can be difficult to combine because both take a lot of time.
5. Use neural networks or other methods to determine how best to satisfy the competing constraints.

The result should be a solution that replaces the balance formulation with a constraint satisfaction solution. Unfortunately, real circumstances and mental limitations can make it hard to identify the specific elements and constraints, so the balance metaphor may be the best that we can do.

Nevertheless, constraint satisfaction provides a general way to think about balance metaphors as well as about literal balancing by the body. Metabalance can be construed as satisfying conflicting constraints between supporting the order of balance and allowing the richness and creativity of imbalance.

My approach to philosophy is strongly opposed to popular trends that are homeostatic in striving to retain equilibrium in ways of thinking. Wittgenstein said that philosophy leaves everything as it is,[21] and much work in analytic philosophy and phenomenology is concerned with analyzing existing concepts rather than developing new ones that are more powerful in explaining the world and improving it. My aim has been to build enriched theories of balance and imbalance that provide scientific explanations of how the world is and that give advice about how to make it better.

Appendix

BALANCE AND IMBALANCE METAPHORS (56 METAPHORS ARRANGED BY TARGET / SOURCE / CHAPTER)

Target	Source	Chapter(s)
Attitudes	Balance, dissonance	8
Autonomic system	Weight scale	7
Balance	Picture puzzle	1, 10
Budget	Weight scale	8
Checks and balances	Weight scale	8
Chemical equation	Algebraic equation	6
Chemical equilibrium	Mechanical equilibrium	6
Climate	Tipping point	6
Cognitive balance and dissonance	Body balance	8
COVID-19 policies	Balance of lives and livelihoods	8
Decision-making	Weight scale	10
Diet	Weight scale	7
Dissonance	Imbalance	8
Doshas	Weight scale	7
Ecological equilibrium	Chemical equilibrium	5

(*continued*)

Target	Source	Chapter(s)
Economic equilibrium	Physical forces equilibrium	8
Electrolytes	Balanced diet	7
Equation	Weight scale	6
Evidence	Weight scale	9
Food and wine	Body balance	8
Hardy-Weinberg equilibrium	Ecological equilibrium	6
Harmony	Balance of consonance and dissonance	9
Health (Chinese)	Balance of yin and yang	7
Heat	Equilibrium	6
Homeostasis	Body balance	6
Hope and despair	Weight scale	9
Humor balance	Weight scale	7
Immune system	Weight scale	7
International payments	Weight scale	8
Justice	Weight scale	8
Love	Falling	8
Meaning of life	Balance	10
Mental illness	Chemical imbalance	8
Morality	Balance	10
Nature	Balance	6
Nausea (mental)	Nausea (physical),	9
Neuroticism	Instability	8
Neurotransmitters	Weight scale,	6

Target	Source	Chapter(s)
Painting	Balance	9
Personality trait	Stability/instability	8
Philosophical thought	Reflective equilibrium	10
Photograph	Balance	9
Political decisions	Weight scale	8
Political equilibrium	Balanced forces	8
Portfolio (economic)	Weight scale	8
Power balance	Weight scale	8
Reflective equilibrium	Economic equilibrium	10
Reporting	Weight scale	8
Shakiness	Fiddler on a roof	9
Social change	Tipping points	8
Theory choice	Balance	9
Thermodynamics	Equilibrium	6
Tipping point	Body balance	6
Vertigo (relationships)	Vertigo (physical),	9
Work-life balance	Body balance	8, 10
Yin-yang balance	Weight scale	6

NOTES

1. BALANCING BODIES AND LIVES

1. Paul Thagard, "Why Does Tai Chi Feel Good?," *Hot Thought* (blog), *Psychology Today*, January 13, 2020, https://www.psychologytoday.com/ca/blog/hot-thought/202001/why-does-tai-chi-feel-good.
2. "Life Is Like Riding a Bicycle. To Keep Your Balance You Must Keep Moving," *Quote Investigator*, June 28, 2015, https://quoteinvestigator.com/2015/06/28/bicycle/.
3. Susan Sontag, *Illness as Metaphor* (New York: Vintage, 1979).
4. In psychology, the term "sensemaking" concerns how people give meaning to their collective experiences, which is a special case of my sensemaking as multimodal parallel constraint satisfaction. The study of problem solving as parallel constraint satisfaction originated with David Marr and Tomaso Poggio, "Cooperative Computation of Stereo Disparity," *Science* 194 (1976): 283–87.

2. BALANCE AND THE BRAIN

1. Karl M. Petruso, "Early Weights and Weighing in Egypt and the Indus Valley," *M Bulletin (Museum of Fine Arts, Boston)* 79 (1981): 44–51.
2. According to Google Scholar, in 2019–2020 more than 350,000 publications had titles that mentioned mechanisms. For analysis of mechanistic explanations, see William Bechtel, *Mental Mechanisms: Philosophical Perspectives on Cognitive Neuroscience* (New York: Routledge, 2008);

Carl F. Craver and Lindley Darden, *In Search of Mechanisms: Discoveries Across the Life Sciences* (Chicago: University of Chicago Press, 2013); Olaf Dammann, *Etiological Explanations* (Boca Raton, FL: CRC, 2020); Stuart Glennan, *The New Mechanical Philosophy* (Oxford: Oxford University Press, 2017); and Paul Thagard, *Natural Philosophy: From Social Brains to Knowledge, Reality, Morality, and Beauty* (New York: Oxford University Press, 2019).

3. My account of balance is based primarily on Jay M. Goldberg, Victor J. Wilson, Dora E. Angelaki, Kathleen E. Cullen, and Kikuro Fukushima, *The Vestibular System: A Sixth Sense* (Oxford: Oxford University Press, 2012). Also useful were the following: Dora E. Angelaki and Kathleen E. Cullen, "Vestibular System: The Many Facets of a Multimodal Sense," *Annual Review of Neuroscience* 31 (2008): 125–50; Kathleen E. Cullen, "The Vestibular System: Multimodal Integration and Encoding of Self-Motion for Motor Control," *Trends in Neurosciences* 35, no. 3 (2012): 185–96; Marianne Dieterich and Thomas Brandt, "Functional Brain Imaging of Peripheral and Central Vestibular Disorders," *Brain* 131, no. 10 (2008): 2538–52; Joseph M. Furman and Thomas Lempert, *Neuro-Otology* (Amsterdam: Elsevier, 2016); John E. Misiaszek, "Neural Control of Walking Balance: If Falling Then React Else Continue," *Exercise and Sport Sciences Reviews* 34, no. 3 (2006): 128–34; and Katherine Sanchez and Fiona J. Rowe, "Role of Neural Integrators in Oculomotor Systems: A Systematic Narrative Literature Review," *Acta Ophthalmologica* 96, no. 2 (2018): e111–e18.
4. Joseph C. Burns and Jennifer S. Stone, "Development and Regeneration of Vestibular Hair Cells in Mammals," *Seminars in Cell and Developmental Biology* 65 (2017): 96–105; W. M. Roberts, J. Howard, and A. J. Hudspeth, "Hair Cells: Transduction, Tuning, and Transmission in the Inner Ear," *Annual Review of Cell Biology* 4, no. 1 (1988): 63–92.
5. Apple's fall detection function is explained at Rob Verger, "The Apple Watch Learned to Detect Falls Using Data from Real Human Mishaps," *Popular Science*, October 3, 2018, https://www.popsci.com/apple-watch-fall-detection/.
6. Goldberg et al., *The Vestibular System*, chap. 5; Helmet T. Karim, Patrick J. Sparto, Howard J. Aizenstein, Joseph M. Furman, Theodore J. Huppert, Kirk I. Erickson, and Patrick J. Loughlin, "Functional MR Imaging of a Simulated Balance Task," *Brain Research* 1555 (2014): 20–27.

7. Andrea M. Green and Dora E. Angelaki, "Internal Models and Neural Computation in the Vestibular System," *Experimental Brain Research* 200, nos. 3–4 (2010): 197–222. More arguments against probabilistic models of the brain are in Thagard, *Natural Philosophy*. For an analysis of the vestibular system using mathematical principles of neural engineering, see Chris Eliasmith and Charles H. Anderson, *Neural Engineering: Computation, Representation, and Dynamics in Neurobiological Systems* (Cambridge, MA: MIT Press, 2003).
8. The following sources discuss constraint satisfaction as a psychological and neural mechanism: James L. McClelland, Daniel Mirman, Donald J. Bolger, and Pranav Khaitan, "Interactive Activation and Mutual Constraint Satisfaction in Perception and Cognition," *Cognitive Science* 38, no. 6 (2014): 1139–89; Stephen J. Read, Eric J. Vanman, and Lynn C. Miller, "Connectionism, Parallel Constraint Satisfaction, and Gestalt Principles: (Re)Introducing Cognitive Dynamics to Social Psychology," *Personality and Social Psychology Review* 1 (1997): 26–53.
9. Patricia T. Alpert, Sally K. Miller, Harvey Wallmann, Richard Havey, Chad Cross, Theresa Chevalia, Carrie B. Gillis, and Keshavan Kodandapari, "The Effect of Modified Jazz Dance on Balance, Cognition, and Mood in Older Adults," *Journal of the American Academy of Nurse Practitioners* 21, no. 2 (2009): 108–15; Con Hrysomallis, "Balance Ability and Athletic Performance," *Sports Medicine* 41, no. 3 (2011): 221–32; Kimberley Hutt and Emma Redding, "The Effect of an Eyes-Closed Dance-Specific Training Program on Dynamic Balance in Elite Pre-Professional Ballet Dancers: A Randomized Controlled Pilot Study," *Journal of Dance Medicine and Science* 18, no. 1 (2014): 3–11; Gabrile Wulf, "Attentional Focus Effects in Balance Acrobats," *Research Quarterly for Exercise and Sport* 79, no. 3 (2008): 319–25.
10. A special issue of *Frontiers in Integrative Neuroscience* in 2016, edited by S. Besnard and others, contains twenty-six articles on the vestibular system in cognitive and memory processes.

3. VERTIGO, NAUSEA, AND FALLS

1. Paul Thagard, *How Scientists Explain Disease* (Princeton, NJ: Princeton University Press, 1999); Lindley Darden, Lipika R. Pal, Kunal Kundu, and John Moult, "The Product Guides the Process: Discovering

Disease Mechanisms," in *Building Theories: Heuristics and Hypotheses in the Sciences*, ed. David Danks and Emiliano Ippoliti (Cham, Switzerland: Springer, 2018), 101–17; Dominic Murphy, "Concepts of Disease and Health," *Stanford Encyclopedia of Philosophy*, 2020, https://plato.stanford.edu/entries/health-disease/; Veli-Pekka Parkkinen, Christian Wallmann, Michael Wilde, Brendan Clarke, Phyllis Illari, Michael P. Kelly, Charles Norell, et al., *Evaluating Evidence of Mechanisms in Medicine: Principles and Procedures* (Berlin: Springer Nature, 2018).

2. Robert E. Post and Lori M. Dickerson, "Dizziness: A Diagnostic Approach," *American Family Physician* 82, no. 4 (2010): 361–68; Alexandre Bisdorff, Gilles Bosser, René Gueguen, and Philippe Perrin, "The Epidemiology of Vertigo, Dizziness, and Unsteadiness and Its Links to Co-Morbidities," *Frontiers in Neurology* 4 (2013): 29.
3. Thomas Brandt, *Vertigo: Its Multisensory Syndromes* (London: Springer, 2003); Joseph M. Furman and Thomas Lempert, *Neuro-Otology* (Amsterdam: Elsevier, 2016); Jay M. Goldberg, Victor J. Wilson, Dora E. Angelaki, Kathleen E. Cullen, and Kikuro Fukushima, *The Vestibular System: A Sixth Sense* (Oxford: Oxford University Press, 2012); Timothy L. Thompson and Ronald Amedee, "Vertigo: A Review of Common Peripheral and Central Vestibular Disorders," *Ochsner Journal* 9, no. 1 (2009): 20.
4. Jack J. Wazen and Deborah Mitchell, *Dizzy: What You Need to Know About Managing and Treating Balance Disorders* (New York: Simon and Schuster, 2008), 19.
5. Lorne S. Parnes, Sumit K. Agrawal, and Jason Atlas, "Diagnosis and Management of Benign Paroxysmal Positional Vertigo (Bppv)," *Canadian Medical Association Journal* 169, no. 7 (2003): 681–93; Timothy C. Hain, Todd M. Squires, and Howard A. Stone, "Clinical Implications of a Mathematical Model of Benign Paroxysmal Positional Vertigo," *Annals of the New York Academy of Sciences* 1039 (2005): 384–94.
6. Ola Alsalman, Jan Ost, Robby Vanspauwen, Catherine Blaivie, Dirk De Ridder, and Sven Vanneste, "The Neural Correlates of Chronic Symptoms of Vertigo Proneness in Humans," *PLoS One* 11, no. 4 (2016): e0152309; Marianne Dieterich and Thomas Brandt, "Functional Brain Imaging of Peripheral and Central Vestibular Disorders," *Brain* 131, no. 10 (2008): 2538–52.

7. Berta Puig, Sandra Brenna, and Tim Magnus, "Molecular Communication of a Dying Neuron in Stroke," *International Journal of Molecular Sciences* 19, no. 9 (2018).
8. Shinichi Iwasaki and Tatsuya Yamasoba, "Dizziness and Imbalance in the Elderly: Age-Related Decline in the Vestibular System," *Aging and Disease* 6, no. 1 (2015): 38–47, 40.
9. Steven D. Rauch, Luis Velazquez-Villaseñor, Paul S. Dimitri, and Saumil N. Merchant, "Decreasing Hair Cell Counts in Aging Humans," *Annals of the New York Academy of Sciences* 942, no. 1 (2001): 220–27.
10. Alexander Borst and Thomas Euler, "Seeing Things in Motion: Models, Circuits, and Mechanisms," *Neuron* 71, no. 6 (2011): 974–94; Eric R. Kandel, James H. Schwartz, Thomas M. Jessell, Steven Siegelbaum, and A. J. Hudspeth, *Principles of Neural Science*, 5th ed. (New York: McGraw-Hill, 2012).
11. More than one hundred visual illusions can be viewed at Michael Bach, "144 Optical Illusions and Visual Phenomena," https://michaelbach.de/ot/.
12. Caroline Tilikete and Alain Vighetto, "Oscillopsia: Causes and Management," *Current Opinion in Neurology* 24, no. 1 (2011): 38–43, 41–42.
13. Richard T. Born and David C. Bradley, "Structure and Function of Visual Area MT," *Annual Review of Neuroscience* 28 (2005): 157–89; Frank Bremmer, Michael Kubischik, Martin Pekel, Klaus-Peter Hoffmann, and Markus Lappe, "Visual Selectivity for Heading in Monkey Area MST," *Experimental Brain Research* 200, no. 1 (2010): 51–60; Syed A. Chowdhury, Katsumasa Takahashi, Gregory C. DeAngelis, and Dora E. Angelaki, "Does the Middle Temporal Area Carry Vestibular Signals Related to Self-Motion?," *Journal of Neuroscience* 29, no. 38 (2009): 12020–30; Uwe J. Ilg, "The Role of Areas MT and MST in Coding of Visual Motion Underlying the Execution of Smooth Pursuit," *Vision Research* 48, no. 20 (2008): 2062–69; Christophe Lopez and Olaf Blanke, "The Thalamocortical Vestibular System in Animals and Humans," *Brain Research Reviews* 67, nos. 1–2 (2011): 119–46; M. James Nichols and William T. Newsome, "Middle Temporal Visual Area Microstimulation Influences Veridical Judgments of Motion Direction," *Journal of Neuroscience* 22, no. 21 (2002): 9530–40; Ben S. Webb, Timothy Ledgeway, and Francesca Rocchi, "Neural Computations Governing

Spatiotemporal Pooling of Visual Motion Signals in Humans," *Journal of Neuroscience* 31, no. 13 (2011): 4917–25; Richard Langton Gregory, "Perceptions as Hypotheses," *Philosophical Transactions of the Royal Society of London B, Biological Sciences* 290, no. 1038 (1980): 181–97. My conjecture that V5 contributes to world-spinning vertigo has no direct evidence, but I have not found alternative explanations.

14. Bill J. Yates, Michael F. Catanzaro, Daniel J. Miller, and Andrew A. McCall, "Integration of Vestibular and Emetic Gastrointestinal Signals That Produce Nausea and Vomiting: Potential Contributions to Motion Sickness," *Experimental Brain Research* 232, no. 8 (2014): 2455–69.
15. Paul W. Glimcher, "Understanding Dopamine and Reinforcement Learning: The Dopamine Reward Prediction Error Hypothesis," *Proceedings of the National Academy of Sciences* 108, no. Supplement 3 (2011): S15647–54.
16. Irene Gazquez and Jose A. Lopez-Escamez, "Genetics of Recurrent Vertigo and Vestibular Disorders," *Current Genomics* 12, no. 6 (2011): 443–50; Lidia Frejo, Ina Giegling, Robert Teggi, Jose A. Lopez-Escamez, and Dan Rujescu, "Genetics of Vestibular Disorders: Pathophysiological Insights," *Journal of Neurology, Neurosurgery & Psychiatry* 263, Suppl 1 (2016): S45–53; Kun-Ling Tsai, Chia-To Wang, Chia-Hua Kuo, Yuan-Yang Cheng, Hsin-I Ma, Ching-Hsia Hung, Yi-Ju Tsai, and Chung-Lan Kao, "The Potential Role of Epigenetic Modulations in Bppv Maneuver Exercises," *Oncotarget* 7, no. 24 (2016): 35522–34; Joaquín Guerra and Ramón Cacabelos, "Pharmacoepigenetics of Vertigo and Related Vestibular Syndromes," in *Pharmacoepigenetics*, ed. Ramón Cacabelos (Cambridge, MA: Academic, 2019), 755–79; Mattieu P. Boisgontier, Boris Cheval, Sima Chalavi, Peter van Ruitenbeek, Inge Leunissen, Oron Levin, Alice Nieuwboer, and Stephan P. Swinnen, "Individual Differences in Brainstem and Basal Ganglia Structure Predict Postural Control and Balance Loss in Young and Older Adults," *Neurobiology of Aging* 50 (2017): 47–59.

4. CONSCIOUSNESS

1. Paul Thagard, *Brain-Mind: From Neurons to Consciousness and Creativity* (Oxford: Oxford University Press, 2019), 160. The three-aspect analysis of concepts is based on Peter Blouw, Eugene Solodkin, Paul Thagard,

and Chris Eliasmith, "Concepts as Semantic Pointers: A Framework and Computational Model," *Cognitive Science* 40 (2016): 1128–62.

2. Francis Crick, *The Astonishing Hypothesis: The Scientific Search for the Soul* (London: Simon and Schuster, 1994); Christof Koch, *The Feeling of Life Itself: Why Consciousness Is Widespread but Can't Be Computed* (Cambridge, MA: MIT Press, 2019); Giulio Tononi, Melanie Boly, Marcello Massimini, and Christof Koch, "Integrated Information Theory: From Consciousness to Its Physical Substrate," *Nature Reviews Neuroscience* 17, no. 7 (2016): 450–61. More critiques of information integration theory are in the following: Paul Thagard and Terrence C. Stewart, "Two Theories of Consciousness: Semantic Pointer Competition vs. Information Integration," *Consciousness and Cognition* 30 (2014): 73–90; Thagard, *Brain-Mind*; Paul Thagard, *Natural Philosophy: From Social Brains to Knowledge, Reality, Morality, and Beauty* (Oxford: Oxford University Press, 2019). For evidence that the brainstem is also important for consciousness, see Bjorn Merker, "Consciousness Without a Cerebral Cortex: A Challenge for Neuroscience and Medicine," *Behavioral and Brain Sciences* 30, no. 1 (2007): 63–81; discussion 81–134.
3. Koch, *The Feeling of Life Itself*, 79.
4. Paul Thagard, *Bots and Beasts: What Makes Machines, Animals, and People Smart?* (Cambridge, MA: MIT Press, 2021).
5. Anthony Damasio and Gil B. Carvalho, "The Nature of Feelings: Evolutionary and Neurobiological Origins," *Nature Reviews Neuroscience* 14, no. 2 (2013): 143–52.
6. Stanislas Dehaene, *Consciousness and the Brain: Deciphering How the Brain Codes Our Thoughts* (New York: Viking, 2014); Stanislas Dehaene, Hakwan Lau, and Sid Kouider, "What Is Consciousness, and Could Machines Have It?," *Science* 358, no. 6362 (2017): 486–92; Olivia Carter, Jakob Hohwy, Jeroen Van Boxtel, Victor Lamme, Ned Block, Christof Koch, and Naotsugu Tsuchiya, "Conscious Machines: Defining Questions," *Science* 359, no. 6374 (2018): 400.
7. Thagard and Stewart, "Two Theories of Consciousness"; Thagard, *Brain-Mind*; Thagard, *Natural Philosophy*; Thagard, *Bots and Beasts*.
8. Chris Eliasmith, *How to Build a Brain: A Neural Architecture for Biological Cognition* (Oxford: Oxford University Press, 2013); Eric Crawford, Matthew Gingerich, and Chris Eliasmith, "Biologically Plausible, Human-Scale Knowledge Representation," *Cognitive Science* 40 (2016):

782–821. Binding can be carried out by convolution and other mathematical operations, providing an effective alternative method of integration beyond Tononi's Φ.

9. George A. Miller, "The Magical Number Seven, Plus or Minus Two: Some Limits on Our Capacity for Processing Information," *Psychological Review* 63 (1956): 81–97. Accounts of attention as interactive competition include Edward E. Smith and Stephen M. Kosslyn, *Cognitive Psychology: Mind and Brain* (Upper Saddle River, NJ: Pearson Prentice Hall, 2007).
10. Philip Ball, "Neuroscience Readies for a Showdown over Consciousness Ideas," *Quanta Magazine*, 2019, https://www.quantamagazine.org/neuroscience-readies-for-a-showdown-over-consciousness-ideas-20190306/; Lucia Melloni, Liad Mudrik, Michael Pitts, and Christof Koch, "Making the Hard Problem of Consciousness Easier," *Science* 372, no. 6545 (2021): 911–12.
11. Max B. Kelz and George A. Mashour, "The Biology of General Anesthesia from Paramecium to Primate," *Current Biology* 29, no. 22 (2019): R1199–210; Zirui Huang, Jun Zhang, Jinsong Wu, George A. Mashour, and Anthony G. Hudetz, "Temporal Circuit of Macroscale Dynamic Brain Activity Supports Human Consciousness," *Science Advances* 6, no. 11 (2020): eaaz0087; James A. Brissenden, Emily J. Levin, David E. Osher, Mark A. Halko, and David C. Somers, "Functional Evidence for a Cerebellar Node of the Dorsal Attention Network," *Journal of Neuroscience* 36, no. 22 (2016): 6083–96; Marcus E. Raichle, "The Brain's Default Mode Network," *Annual Review of Neuroscience* 38 (2015): 433–47.
12. Aron K. Barbey, "Network Neuroscience Theory of Human Intelligence," *Trends in Cognitive Sciences* 22, no. 1 (2018): 8–20; Lisa F. Barrett and Ajay B. Satpute, "Large-Scale Brain Networks in Affective and Social Neuroscience: Towards an Integrative Functional Architecture of the Brain," *Current Opinion in Neurobiology* 23, no. 3 (2013): 361–72; Roger E. Beaty, Mathias Benedek, Scott Barry Kaufman, and Paul J. Silvia, "Default and Executive Network Coupling Supports Creative Idea Production," *Scientific Reports* 5 (2015): 10964; Randy L. Buckner and Fenna M. Krienen, "The Evolution of Distributed Association Networks in the Human Brain," *Trends in Cognitive Sciences* 17, no. 12 (2013): 648–65; Yonghui Li, Yong Liu, Jun Li, Wen Qin, Kuncheng Li, Chunsui Yu, and Tinazi Jiang, "Brain Anatomical Network and Intelligence,"

PLoS Computational Biology 5, no. 5 (May 2009): e1000395; George A. Mashour and Anthony G. Hudetz, "Neural Correlates of Unconsciousness in Large-Scale Brain Networks," *Trends in Neurosciences* 41, no. 3 (2018): 150–60; Olaf Sporns, "The Human Connectome: Origins and Challenges," *Neuroimage* 80 (2013): 53–61.

13. Alexandre Bisdorff, Michael Von Brevern, Thomas Lempert, and David E. Newman-Toker, "Classification of Vestibular Symptoms: Towards an International Classification of Vestibular Disorders," *Journal of Vestibular Research* 19, nos. 1–2 (2009): 1–13; Christophe Lopez, "A Neuroscientific Account of How Vestibular Disorders Impair Bodily Self-Consciousness," *Frontiers in Integrative Neuroscience* 7 (2013): 91; Christian Pfeiffer, Andrew Serino, and Olaf Blanke, "The Vestibular System: A Spatial Reference for Bodily Self-Consciousness," *Frontiers in Integrative Neuroscience* 8 (2014): 31.
14. Bisdorff et al., "Classification of Vestibular Symptoms," 7.
15. Bisdorff et al., "Classification of Vestibular Symptoms," 9.
16. Ivana Kajić, Tobias C. Schröder, Terrence C. Stewart, and Paul Thagard, "The Semantic Pointer Theory of Emotions," *Cognitive Systems Research* 58 (2019): 35–53; Paul Thagard, Laurette Larocque, and Ivana Kajić, "Emotional Change: Neural Mechanisms Based on Semantic Pointers," *Emotion* (forthcoming); Archana Rajagopalan, K. V. Jinu, Kumar Sai Sailesh, Soumya Mishra, Udaya Kumar Reddy, and Joseph Kurien Mukkadan, "Understanding the Links Between Vestibular and Limbic Systems Regulating Emotions," *Journal of Natural Science, Biology and Medicine* 8, no. 1 (2017): 11–15.
17. Applications of semantic pointers to language are discussed in Eliasmith, *How to Build a Brain*, and Thagard, *Brain-Mind*.
18. The hard problem of consciousness can't be solved: David J. Chalmers, *The Conscious Mind* (Oxford: Oxford University Press, 1996). Yes it can: Thagard, *Brain-Mind*; Thagard, *Natural Philosophy*.
19. Paul Thagard, "Explanatory Identities and Conceptual Change," *Science and Education* 23 (2014): 1531–48; Paul Thagard, "Thought Experiments Considered Harmful," *Perspectives on Science* 22 (2014): 288–305.
20. Howard Robinson, "Dualism," *Stanford Encyclopedia of Philosophy*, 2016, https://plato.stanford.edu/entries/dualism/; Paul Thagard, *The Brain and the Meaning of Life* (Princeton, NJ: Princeton University Press, 2010); Thagard, *Natural Philosophy*.

21. Philip Goff, William Seager, and Sean Allen-Hermanson, "Panpsychism," *Stanford Encyclopedia of Philosophy*, 2017, https://plato.stanford.edu/entries/panpsychism/; Irene Tracey and Patrick W. Mantyh, "The Cerebral Signature for Pain Perception and Its Modulation," *Neuron* 55, no. 3 (2007): 377–91.
22. Paul Thagard, "Energy Requirements Undermine Substrate Independence and Mind-Body Functionalism," *Philosophy of Science* (forthcoming); Thagard, *Bots and Beasts*.

5. HOW METAPHORS WORK

1. Keith J. Holyoak, *The Spider's Thread: Metaphor in Mind, Brain, and Poetry* (Cambridge, MA: MIT Press, 2019). On metaphor and analogy, see also the following: Gilles Fauconnier and Mark Turner, *The Way We Think* (New York: Basic, 2002); Ray W. Gibbs, *Metaphor Wars* (Cambridge: Cambridge University Press, 2017); David Hills, "Metaphor," *Stanford Encyclopedia of Philosophy*, 2016, https://plato.stanford.edu/entries/metaphor/; Keith J. Holyoak, "Analogy and Relational Reasoning," in *The Oxford Handbook of Thinking and Reasoning*, ed. Keith J. Holyoak and Robert G. Morrison (Oxford: Oxford University Press, 2012), 234–59; Keith J. Holyoak and Paul Thagard, *Mental Leaps: Analogy in Creative Thought* (Cambridge, MA: MIT Press, 1995); Andrew Ortony, ed., *Metaphor and Thought* (Cambridge: Cambridge University Press, 1993); George Lakoff and Mark Johnson, *Metaphors We Live By* (Chicago: University of Chicago Press, 1980); Mark J. Landau, *Conceptual Metaphor in Social Psychology: The Poetics of Everyday Life* (New York: Routledge, 2017); Paul Thagard, *Brain-Mind: From Neurons to Consciousness and Creativity* (Oxford: Oxford University Press, 2019); Paul H. Thibodeau, Rose K. Hendricks, and Lera Boroditsky, "How Linguistic Metaphor Scaffolds Reasoning," *Trends in Cognitive Sciences* 21, no. 11 (2017): 852–63.
2. Mark Johnson, *The Body in the Mind* (Chicago: University of Chicago Press, 1987).
3. Lawrence W. Barsalou, "Perceptual Symbol Systems," *Behavioral and Brain Sciences* 22 (1999): 577–660; Lawrence W. Barsalou, "Simulation, Situated Conceptualization, and Prediction," *Philosophical Transactions of the Royal Society of London B, Biological Sciences* 364 (2009): 1218–89;

Robert A. Mason and Marcel Adam Just, "Neural Representations of Procedural Knowledge," *Psychological Science* 31, no. 6 (2020): 729–40.

4. Holyoak and Thagard, *Mental Leaps*.
5. Brian F. Bowdle and Dedre Gentner, "The Career of Metaphor," *Psychological Review* 112, no. 1 (2005): 193–216; Dedre Gentner and Brian Bowdle, "Metaphor as Structure-Mapping," in *The Cambridge Handbook of Metaphor and Thought*, ed. R. W. Gibbs (Cambridge: Cambridge University Press, 2008), 109–28; Eileen R. Cardillo, Christine E. Watson, Gwenda L Schmidt, Alexander Kranjec, and Anjan Chatterjee, "From Novel to Familiar: Tuning the Brain for Metaphors," *Neuroimage* 59, no. 4 (2012): 3212–21.
6. Paul Thagard, *Mind-Society: From Brains to Social Sciences and Professions* (New York: Oxford University Press, 2019). For a list of more than twenty publications using cognitive-affective maps, go to https://paulthagard.com/links/cognitive-affective-maps/.
7. Lakoff and Johnson, *Metaphors We Live By*; George Lakoff and Mark Johnson, *Philosophy in the Flesh: The Embodied Mind and Its Challenge to Western Thought* (New York: Basic Books, 1999); Randy Allen Harris, *The Linguistic Wars*, 2nd ed. (Oxford: Oxford University Press, 2021).
8. Lakoff and Johnson, *Philosophy in the Flesh*, 185.
9. Lakoff and Johnson, *Philosophy in the Flesh*, 184.
10. Paul Thagard, *Natural Philosophy: From Social Brains to Knowledge, Reality, Morality, and Beauty* (New York: Oxford University Press, 2019), chap. 4.
11. Paul Thagard, *The Cognitive Science of Science: Explanation, Discovery, and Conceptual Change* (Cambridge, MA: MIT Press, 2012). A defense of the objectivity of scientific revolutions is here: Paul Thagard, *Conceptual Revolutions* (Princeton, NJ: Princeton University Press, 1992).
12. Gerard J. Steen, Aletta G. Dorst, J. Berenike Herrmann, Anna A. Kaal, and Tina Krennmayr, "Metaphor in Usage," *Cognitive Linguistics* 21, no. 4 (2010): 765–96.
13. Thagard, *Natural Philosophy*, chap. 4.
14. Extreme embodiment views include the following: Anthony Chemero, *Radical Embodied Cognitive Science* (Cambridge, MA: MIT Press, 2009); Hubert L. Dreyfus, "Why Heideggerian AI Failed and How Fixing It Would Require Making It More Heideggerian," *Philosophical Psychology* 20 (2007): 247–68; Shaun Gallagher, *How the Body Shapes the*

Mind (Oxford: Oxford University Press, 2006). Moderate embodiment views include the following: Barsalou, "Perceptual Symbol Systems"; Ray W. Gibbs, *Embodiment and Cognitive Science* (Cambridge: Cambridge University Press, 2005); Lorenzo Magnani, *Abductive Cognition: The Epistemological and Eco-Cognitive Dimensions of Hypothetical Reasoning* (Berlin: Springer, 2009); Thagard, *The Cognitive Science of Science*; Thagard, *Brain-Mind.*

15. Thagard, *Natural Philosophy*, chap. 5.
16. Paul Thagard, *Bots and Beasts: What Makes Machines, Animals, and People Smart?* (Cambridge, MA: MIT Press, 2021), chap. 5.

6. NATURE

1. James Franklin, *An Aristotelian Realist Philosophy of Mathematics* (Houndmills, UK: Palgrave Macmillan, 2014); Paul Thagard, *Natural Philosophy: From Social Brains to Knowledge, Reality, Morality, and Beauty* (New York: Oxford University Press, 2019), chap. 10.
2. George Lakoff and Rafael E. Núñez, *Where Mathematics Comes From: How the Embodied Mind Brings Mathematics into Being* (New York: Basic, 2000).
3. Jürgen Renn and Peter Damerow, *The Equilibrium Controversy: Guidobaldo Del Monte's Critical Notes on the Mechanics of Jordanus and Benedetti and Their Historical and Conceptual Backgrounds* (Berlin: Edition Open Access, 2017).
4. The idea of a metaphorical chain reaction derives from the description by Holyoak and Thagard (Keith Holyoak and Paul Thagard, *Mental Leaps: Analogy in Creative Thought* [Cambridge, MA: MIT Press], 243) of an analogical chain reaction in their own work.
5. On the analogical discovery of the wave theory of sound, see Paul Thagard, *Computational Philosophy of Science* (Cambridge, MA: MIT Press, 1988).
6. Maurice W. Lindauer, "The Evolution of the Concept of Chemical Equilibrium from 1775 to 1923," *Journal of Chemical Education* 39, no. 8 (1962): 384.
7. Frank N. Egerton, "Changing Concepts of the Balance of Nature," *Quarterly Review of Biology* 48, no. 2 (1973): 322–50; David Suzuki, *The Sacred Balance: Rediscovering Our Place in Nature*, updated and expanded (Vancouver, BC: Greystone, 2007).

8. Elizabeth Kolbert, *The Sixth Extinction: An Unnatural History* (London: A&C Black, 2014); Frederik Saltre and Corey J. A. Bradshaw, "Are We Really in a 6th Mass Extinction? Here's the Science," *Science Alert*, November 18, 2019, https://www.sciencealert.com/here-s-how-biodiversity-experts-recognise-that-we-re-midst-a-mass-extinction; Paul Thagard, *Bots and Beasts: What Makes Machines, Animals, and People Smart?* (Cambridge, MA: MIT Press, 2021), chap. 7, discusses why extinction is bad.
9. John Kricher, *The Balance of Nature: Ecology's Enduring Myth* (Princeton, NJ: Princeton University Press, 2009), 19. Other critics of the *balance of nature* metaphor include the following: Gregory Cooper, "Must There Be a Balance of Nature?," *Biology and Philosophy* 16, no. 4 (2001): 481–506; Kim Cuddington, "The 'Balance of Nature' Metaphor and Equilibrium in Population Ecology," *Biology and Philosophy* 16, no. 4 (2001): 463–79; Dennis E. Jelinski, "There Is No Mother Nature—There Is No Balance of Nature: Culture, Ecology and Conservation," *Human Ecology* 33, no. 2 (2005): 271–88. Other dubious biological metaphors include *mother nature* and the *circle of life*.
10. Stuart L. Pimm, *The Balance of Nature: Ecological Issues in the Conservation of Species and Communities* (Chicago: University of Chicago Press, 1991).
11. Stephen Jay Gould and Niles Eldredge, "Punctuated Equilibrium Comes of Age," *Nature* 366, no. 6452 (1993): 223–27.
12. Malcolm Gladwell, *The Tipping Point: How Little Things Can Make a Big Difference* (New York: Little, Brown, 2000).
13. Manjana Milkoreit, Jennifer Hodbod, Jacopo Baggio, Karina Benessaiah, Rafael Calderón-Contreras, Jonathan F. Donges, Jean-Denis Mathias, et al., "Defining Tipping Points for Social-Ecological Systems Scholarship—An Interdisciplinary Literature Review," *Environmental Research Letters* 13, no. 3 (2018): 9.
14. Timothy M. Lenton, Johan Rockström, Owen Gaffney, Stefan Rahmstorf, Katherine Richardson, Will Steffen, and Hans Joachim Schellnhuber, "Climate Tipping Points—Too Risky to Bet Against," *Nature* 575 (2019): 592–95; Marten Scheffer, *Critical Transitions in Nature and Society* (Princeton, NJ: Princeton University Press, 2009).
15. Michael Ranney, Daniel Reinholz, and Lloyd Goldwasser, "How Does Climate Change ('Global Warming') Work? The Mechanism of Global Warming, an Extra Greenhouse Effect," https://www.howglobalwarmingworks.org/35-words.html.

16. Michael A. Ranney and Dav Clark, "Climate Change Conceptual Change: Scientific Information Can Transform Attitudes," *Topics in Cognitive Science* 8, no. 1 (2016): 49–75.

7. MEDICINE

1. Lawrence I. Conrad, Michael Neve, Vivian Nutton, Roy Porter, and Andrew Wear, *The Western Medical Tradition: 800 BC to AD 1800* (Cambridge: Cambridge University Press, 1995); William F. Bynum, Anne Hardy, Stephen Jacyna, Christopher Lawrence, and E. M. Tansey, *The Western Medical Tradition: 1800–2000* (Cambridge: Cambridge University Press, 2006).
2. Hippocrates, *Delphi Complete Works of Hippocrates* (East Sussex, UK: Delphi Classics, 2015), 334.
3. Shen Ziyin and Chen Zelin, *The Basis of Traditional Chinese Medicine* (Boston: Shambhala, 1994); Manfred Porkert and Christian Ullmann, *Chinese Medicine*, trans. M. Howson (New York: Morrow, 1988); Paul Thagard and Jing Zhu, "Acupuncture, Incommensurability, and Conceptual Change," in *Intentional Conceptual Change*, ed. G. M. Sinatra and P. R. Pintrich (Mahwah, NJ: Erlbaum, 2003), 79–102; Ilza Veith, trans., *The Yellow Emperor's Classic of Internal Medicine* (Oakland: University of California Press, 1975).
4. Eric Manheimer, Susan Wieland, Elizabeth Kimbrough, Ker Cheng, and Brian M. Berman, "Evidence from the Cochrane Collaboration for Traditional Chinese Medicine Therapies," *Journal of Alternative and Complementary Medicine* 15, no. 9 (2009): 1001–14.
5. Chiu-Wen Chiu, Tsung-Chieh Lee, Po-Chi Hsu, Chia-Yun Chen, Shun-Chang Chang, John Y. Chiang, and Lun-Chien Lo, "Efficacy and Safety of Acupuncture for Dizziness and Vertigo in Emergency Department: A Pilot Cohort Study," *BMC Complementary and Alternative Medicine* 15 (2015): 173; Edzard Ernst, "Acupuncture: What Does the Most Reliable Evidence Tell Us?," *Journal of Pain Symptom Management* 37, no. 4 (2009): 709–14; Shenbin Liu, Zhi-Fu Wang, Yuang-Shai Su, Russell S. Ray, Xiang-Hong Jing, Yan-Qing Wang, and Quifu Ma, "Somatotopic Organization and Intensity Dependence in Driving Distinct Npy-Expressing Sympathetic Pathways by Electroacupuncture," *Neuron* 108, no. 3 (2020): 436–50.e7; Carol A. Paley and Mark I. Johnson,

"Acupuncture for the Relief of Chronic Pain: A Synthesis of Systematic Reviews," *Medicina* 56 (2019): 1; Gabriel Stux, Brian Berman, and Bruce Pomeranz, *Basics of Acupuncture* (Berlin: Springer, 2012).

6. Manisha Kshirsagar and Ana C. Magno, *Ayurveda: A Quick Reference Handbook* (Twin Lakes, WI: Lotus, 2011); Sahara Rose Ketabi, *Ayurveda* (Indianapolis, IN: Dorling Kindersley, 2017); Bhushan Patwardhan, Dnyaneshwar Warude, Palpu Pushpangadan, and Narendra Bhatt, "Ayurveda and Traditional Chinese Medicine: A Comparative Overview," *Evidence-Based Complementary and Alternative Medicine* 2, no. 4 (2005): 465–73. My critique of Deepak Chopra's latest book is here: Paul Thagard, "Deepak Chopra's Infinite Potential," *Hot Thought* (blog), *Psychology Today*, October 6, 2020, https://www.psychologytoday.com/ca/blog/hot-thought/202010/deepak-chopra-s-infinite-potential.
7. Paul Thagard, *How Scientists Explain Disease* (Princeton, NJ: Princeton University Press, 2000). See also note 1 to chapter 3.
8. Paul Thagard, "Thought Experiments Considered Harmful," *Perspectives on Science* 22, no. 2 (2014): 122–39; Paul Thagard, *Natural Philosophy: From Social Brains to Knowledge, Reality, Morality, and Beauty* (New York: Oxford University Press, 2019), chap. 3.
9. Manheimer et al., "Evidence from the Cochrane Collaboration."
10. Olaf Dammann, Ted Poston, and Paul Thagard, "How Do Medical Researchers Make Causal Inferences?," in *What Is Scientific Knowledge? An Introduction to Contemporary Epistemology of Science*, ed. Kevin McCain and Kostas Kampourakis (New York: Routledge, 2019), 33–51.
11. Paul Thagard, "Evolution, Creation, and the Philosophy of Science," in *Evolution, Epistemology, and Science Education*, ed. Roger Taylor and Michel Ferrari (Milton Park, UK: Routledge, 2010), 20–37.
12. Ziva Kunda, "The Case for Motivated Reasoning," *Psychological Bulletin* 108 (1990): 480–98.
13. Tor D. Wager and Lauren Y. Atlas, "The Neuroscience of Placebo Effects: Connecting Context, Learning and Health," *Nature Reviews Neuroscience* 16, no. 7 (2015): 403–18.
14. Daniel J. Levitin, *Successful Aging: A Neuroscientist Explores the Power and Potential of Our Lives* (New York: Dutton, 2020), chap. 10; Charles M. Tipton, "The History of 'Exercise Is Medicine' in Ancient Civilizations," *Advances in Physiology Education* 38, no. 2 (2014): 109–17.

15. Patricia Huston and Bruce McFarlane, "Health Benefits of Tai Chi: What Is the Evidence?," *Canadian Family Physician* 62, no. 11 (2016): 881–90; Chenchen Wang, Raveendhara Bannuru, Judith Ramel, Bruce Kupelnick, Tammy Scott, and Christopher H. Schmid, "Tai Chi on Psychological Well-Being: Systematic Review and Meta-Analysis," *BMC Complementary and Alternative Medicine* 10 (2010): 23; Peter M. Wayne and Mark Fuerst, *The Harvard Medical School Guide to Tai Chi: 12 Weeks to a Healthy Body, Strong Heart, and Sharp Mind* (Boston: Shambhala, 2013); Angus P. Yu, Bjorn T. Tam, Christopher W. Lai, Doris S. Yu, Jean Woo, Ka-Fai Chung, Stanley S. Hui, et al., "Revealing the Neural Mechanisms Underlying the Beneficial Effects of Tai Chi: A Neuroimaging Perspective," *American Journal of Chinese Medicine* 46, no. 2 (2018): 231–59.
16. Randall Collins, *Interaction Ritual Chains* (Princeton, NJ: Princeton University Press, 2004).
17. Pamela E. Jeter, Amélie-Françoise Nkodo, Steffany H. Moonaz, and Gislin Dagnelie, "A Systematic Review of Yoga for Balance in a Healthy Population," *Journal of Alternative and Complementary Medicine* 20, no. 4 (2014): 221–32; Ann-Kathrin Rogge, Brigitte Roder, Astrid Zech, Volker Nagel, Karsten Hollander, Klaus-Michael Braumann, and Kirsten Hotting, "Balance Training Improves Memory and Spatial Cognition in Healthy Adults," *Scientific Reports* 7, no. 1 (2017): 5661; Roderik J. S. Gerritsen and Guido P. H. Band, "Breath of Life: The Respiratory Vagal Stimulation Model of Contemplative Activity," *Frontiers in Human Neuroscience* 12 (2018): 397.
18. NHS (UK National Health Service), "Eat Well," https://www.nhs.uk/live-well/eat-well/.
19. Kimberly Holland, "All About Electrolyte Disorders," Healthline, https://www.healthline.com/health/electrolyte-disorders.
20. Junjiao Wu and Yu Tang, "Revisiting the Immune Balance Theory: A Neurological Insight into the Epidemic of Covid-19 and Its Alike," *Frontiers in Neurology* 11 (2020): 566680.
21. "A Guide to Neurotransmitter Balance," *PowerOnPowerOff* (blog), updated May 13, 2020, https://poweronpoweroff.com/blogs/longform/a-guide-to-neurotransmitter-balance.
22. François M. Abboud, "In Search of Autonomic Balance: The Good, the Bad, and the Ugly," *American Journal of Physiology: Regulatory, Integrative, and Comparative Physiology* 298, no. 6 (2010): R1449–67.

23. "Vaccinate or Not? Pros and Cons of Naturopathy," *KUTV News*, August 1, 2019, http://kutv.com/features/sinclair-cares/vaccinate-or-not-pros-and-cons-of-naturopathy.
24. Arran Stibbe, "The Role of Image Systems in Complementary Medicine," *Complementary Therapies in Medicine* 6, no. 4 (1998): 190–94.

8. SOCIETY

1. S. Max Brown and Tanveer Naseer, *Leadership Vertigo: Why Even the Best Leaders Go Off Course and How They Can Get Back on Track* (Sanger, CA: Familius, 2014).
2. Suniya S. Luthar, Emily L. Lyman, and Elizabeth J. Crossman, "Resilience and Positive Psychology," in *Handbook of Developmental Psychopathology*, ed. M. Lewis and K. D. Rudolph (New York: Springer, 2014), 125–40.
3. David R. Kille, Amanda L. Forest, and Joanne V. Wood, "Tall, Dark, and Stable: Embodiment Motivates Mate Selection Preferences," *Psychological Science* 24, no. 1 (2013): 112–14; Amanda L. Forest, David R. Kille, Joanne V. Wood, and Lindsay R. Stehouwer, "Turbulent Times, Rocky Relationships: Relational Consequences of Experiencing Physical Instability," *Psychological Sciences* 26, no. 8 (2015): 1261–71.
4. Fritz Heider, *The Psychology of Interpersonal Relations* (New York: Wiley, 1958), 198; Fritz Heider, "Attitudes and Cognitive Organization," *Journal of Psychology* 21, no. 1 (1946): 107–12.
5. Leon Festinger, *A Theory of Cognitive Dissonance* (Stanford, CA: Stanford University Press, 1957); Eddie Harmon-Jones, ed., *Cognitive Dissonance: Reexamining a Pivotal Theory in Psychology* (Washington, DC: American Psychological Association, 2019). For the understanding of dissonance as constraint satisfaction, see the following: Stephen J. Read, Eric J. Vanman, and Lynn C. Miller, "Connectionism, Parallel Constraint Satisfaction, and Gestalt Principles," *Personality and Social Psychology Review* 1, no. 1 (1997); Thomas R. Shultz and Mark R. Lepper, "Cognitive Dissonance Reduction as Constraint Satisfaction," *Psychological Review* 103 (1996): 219–40. Discussions of balance, fit, consonance, harmony, and coherence are similar ways of getting at parallel constraint satisfaction: see chapter 10.
6. Festinger, *A Theory of Cognitive Dissonance*, 6–7.

7. David Healy, "Serotonin and Depression," *BMJ* 350 (2015): h1771; Jeffrey R. Lacasse and Jonathan Leo, "Antidepressants and the Chemical Imbalance Theory of Depression: A Reflection and Update on the Discourse," *The Behavior Therapist* 38, no. 7 (2015): 206–13.
8. Alfred Marshall, *Principles of Economics* (London: Macmillan, 1890), 383; Daniel M. Hausman, *The Inexact and Separate Science of Economics* (Cambridge: Cambridge University Press, 1992).
9. Alfred Marshall, *Principles of Economics*, 8th ed. (London: Macmillan, 1920), 345–46.
10. Daniel Kahneman and Amos Tversky, eds., *Choices, Values, and Frames* (Cambridge: Cambridge University Press, 2000).
11. Eric D. Beinhocker, *The Origin of Wealth: Evolution, Complexity, and the Radical Remaking of Economics* (Cambridge, MA: Harvard Business School Press, 2006); Eric D. Beinhocker, J. Doyne Farmer, and Cameron Hepburn, "The Tipped Point: How the G20 Can Lead the Transition to a Prosperous Clean Energy Economy," G20 Insights, May 25, 2018, https://www.g20-insights.org/policy_briefs/the-tipping-point-how-the-g20-can-lead-the-transition-to-a-prosperous-clean-energy-economy/.
12. Paul Thagard, *Mind-Society: From Brains to Social Services and Professions* (Oxford: Oxford University Press, 2019), chap. 7.
13. Joseph A. Schumpeter, *Capitalism, Socialism and Democracy* (New York: Harper, 1942); Clayton M. Christensen, Michael E. Raynor, and Rory McDonald, "What Is Disruptive Innovation?," *Harvard Business Review* 93, no. 12 (2015): 44–53; Joseph E. Stiglitz, *Freefall: America, Free Markets, and the Sinking of the World Economy* (New York: Norton, 2010).
14. Roger L. Martin, *When More Is Not Better: Overcoming America's Obsession with Economic Efficiency* (Boston: Harvard Business Review Press, 2020).
15. "Economic Vertigo Hits World Markets," *Financial Times*, August 19, 2011, https://www.ft.com/content/e437115e-ca5e-11e0-a0dc-00144feabdc0.
16. Choe Sang-Hun, "New Covid-19 Outbreaks Test South Korea's Strategy," *New York Times*, September 2, 2020, https://www.nytimes.com/2020/09/02/world/asia/south-korea-covid-19.html?referringSource=articleShare.
17. Government of Ontario, "Approach to Reopening Schools for the 2020–2021 School Year," Ontario.ca, last updated June 28, 2021, https://www.ontario.ca/page/approach-reopening-schools-2020-2021-school-year.
18. Peter C. Ordeshook and Kenneth A. Shepsle, eds., *Political Equilibrium: A Delicate Balance* (Boston: Kluwer-Nijhoff, 1982); Daniel Sutter,

"The Democratic Efficiency Debate and Definitions of Political Equilibrium," *Review of Austrian Economics* 15, nos. 2–3 (2002): 199–209; J. Eli Margolis, "Understanding Political Stability and Instability," *Civil Wars* 12, no. 3 (2010): 326–45.

19. Herbert Butterfield, "Balance of Power," in *Dictionary of the History of Ideas*, ed. P. P. Wiener (New York: Scribner's, 1973), 179–88, 179.
20. A. J. C. Edwards, *Nuclear Weapons: The Balance of Terror, the Quest for Peace* (Albany: State University of New York Press, 1986).
21. Talcott Parsons, *The Structure of Social Action* (New York: Free Press, 1949).
22. Morton Grodzins, "Metropolitan Segregation," *Scientific American* 197, no. 4 (1957): 33–41.
23. Malcolm Gladwell, *The Tipping Point: How Little Things Can Make a Big Difference* (New York: Little, Brown, 2006).
24. Theodore C. Foin and William G. Davis, "Equilibrium and Nonequilibrium Models in Ecological Anthropology: An Evaluation of 'Stability' in Maring Ecosystems in New Guinea," *American Anthropologist* 89, no. 1 (1987): 9–31.
25. Jared Diamond, *Collapse: How Societies Choose to Fail or Succeed* (New York: Penguin, 2011).
26. Christopher E. Clarke, "A Question of Balance: The Autism-Vaccine Controversy in the British and American Elite Press," *Science Communication* 30, no. 1 (2008): 77–107; Mikkel Gerken, "How to Balance Balanced Reporting and Reliable Reporting," *Philosophical Studies* 177, no. 10 (2020): 3117–42; Karin Wahl-Jorgensen, Mike Berry, Iñaki Garcia-Blanco, Lucy Bennett, and Jonathan Cable, "Rethinking Balance and Impartiality in Journalism? How the BBC Attempted and Failed to Change the Paradigm," *Journalism* 18, no. 7 (2017): 781–800.
27. The following are some websites on balanced wine and food: Vicki Denig, "What Is a 'Well-Balanced' Wine and Why Should I Care?," VinePair, September 20, 2016, https://vinepair.com/articles/what-exactly-is-a-well-balanced-wine/; May Gorman-McAdams, "Wine Words: Balance," Kitchn, May 7, 2012, https://www.thekitchn.com/wine-words-balance-170792; "What Is Balance in Beer?," Boak and Bailey, February 15, 2013, https://boakandbailey.com/2013/02/what-is-balance-in-beer/; Madeline Puckette, "6 Essential Basics to Food and Wine Pairing," Wine Folly, April 30, 2012, https://winefolly.com/tips/food-and-wine-pairing/.

9. THE ARTS

1. Rohinton Mistry, *A Fine Balance* (Toronto, ON: McClelland and Stewart, 1995), 268.
2. Mistry, *A Fine Balance*, 652.
3. Jean-Paul Sartre, *Nausea*, trans. L. Alexander (New York: New Directions, 1964).
4. Robert B. Pippin, *The Philosophical Hitchcock: "Vertigo" and the Anxieties of Unknowingness* (Chicago: University of Chicago Press, 2017); Dan Auilier, *Vertigo: The Making of a Hitchcock Classic* (Seattle: Amazon Kindle, 2013).
5. Paul Thagard, *Hot Thought: Mechanisms and Applications of Emotional Cognition* (Cambridge, MA: MIT Press, 2006).
6. Michael Hahn, "Equalization 101: Everything Musicians Need to Know About EQ," LANDR, November 10, 2018, https://blog.landr.com/eq-basics-everything-musicians-need-know-eq/; Camden Shaw, "Balancing the Viola in a String Quartet," *Studio Blog* (blog), American Viola Society, April 15, 2014, http://studio.americanviolasociety.org/studio2/2014/04/15/balance-by-camden-shaw/.
7. Paul Thagard, *Natural Philosophy: From Social Brains to Knowledge, Reality, Morality, and Beauty* (New York: Oxford University Press, 2019), 247.
8. Shelley Esaak, "What Is Balance in Art and Why Does It Matter?," ThoughtCo., January 31, 2020, https://www.thoughtco.com/definition-of-balance-in-art-182423. Four kinds of balance in art (symmetrical, asymmetrical, radial, and crystallographic) are described in "4 Types of Balance in Art and Design and Why You Need Them," Shutterstock, November 30, 2020, https://www.shutterstock.com/blog/types-of-balance-in-art. Architectural balance is discussed in Joffre Essley, "Architectural Balance," House Design Coffee, https://www.house-design-coffee.com/architectural-balance.html.
9. Mark Johnson, *The Body in the Mind* (Chicago: University of Chicago Press, 1987).
10. Rudolf Arnheim, *Art and Visual Perception: The New Version* (Berkeley: University of California Press, 1974).
11. Francis Hutcheson, *An Inquiry into the Original of Our Ideas of Beauty and Virtue: In Two Treatises* (London: R. Ware, 1753).
12. Thagard, *Natural Philosophy*, chap. 9.

10. PHILOSOPHY

1. Paul Thagard and Craig Beam, "Epistemological Metaphors and the Nature of Philosophy," *Metaphilosophy* 35 (2004): 504–16. The crossword puzzle analogy for knowledge is advocated by Susan Haack, *Evidence and Inquiry: Towards Reconstruction in Epistemology* (Oxford: Blackwell, 1993).
2. Alvin Goldman, *Epistemology and Cognition* (Cambridge, MA: Harvard University Press, 1986); Paul Thagard, *Natural Philosophy: From Social Brains to Knowledge, Reality, Morality, and Beauty* (New York: Oxford University Press, 2019), chap. 3.
3. David Hume, *Enquiries Concerning Human Understanding and Concerning the Principles of Morals*, ed. L. A. Selby-Bigge, 2nd ed. (Oxford: Oxford University Press, 1972), section 10.
4. Paul Thagard, "Explanatory Coherence," *Behavioral and Brain Sciences* 12 (1989): 435–67; Paul Thagard, *Conceptual Revolutions* (Princeton, NJ: Princeton University Press, 1992); Paul Thagard, *Coherence in Thought and Action* (Cambridge, MA: MIT Press, 2000); Paul Thagard, *The Cognitive Science of Science: Explanation, Discovery, and Conceptual Change* (Cambridge, MA: MIT Press, 2012); Paul Thagard, "The Cognitive Science of COVID-19: Acceptance, Denial, and Belief Change," *Methods* 195 (2021): 92-102, https://doi.org/10.1016/j.ymeth.2021.03.009.
5. Daniel Politi, "Richard Thaler Wins Economics Nobel for Recognizing People Are Irrational," *Slate*, October 9, 2017, http://www.slate.com/blogs/the_slatest/2017/10/09/richard_thaler_wins_economics_nobel_for_recognizing_that_people_are_irrational.html.
6. Paul Thagard, *The Brain and the Meaning of Life* (Princeton, NJ: Princeton University Press, 2010); Paul Thagard, *Brain-Mind: From Neurons to Consciousness and Creativity* (New York: Oxford University Press, 2019).
7. George Ainslie, "Willpower with and Without Effort," *Behavioral and Brain Sciences* 44 (2021): e30; Esther K. Diekhof and Oliver Gruber, "When Desire Collides with Reason: Functional Interactions Between Anteroventral Prefrontal Cortex and Nucleus Accumbens Underlie the Human Ability to Resist Impulsive Desires," *Journal of Neuroscience* 30, no. 4 (2010): 1488–93; Samuel M. McClure, David I. Laibson, George Loewenstein, and Jonathan D. Cohen, "Separate Neural Systems Value Immediate and Delayed Monetary Rewards," *Science* 306, no. 5695 (2004): 503–7; Paul Thagard, "How Rationality Is Bounded by

the Brain," in *Routledge Handbook of Bounded Rationality*, ed. Riccardo Viale, 398-406 (London: Routledge, 2021); Amos Tversky and Daniel Kahneman, "The Framing of Decisions and the Psychology of Choice," *Science* 211 (1981): 453–58.

8. Wikipedia, s.v. "Precautionary Principle," last modified January 5, 2022, https://en.wikipedia.org/wiki/Precautionary_principle.
9. Thagard, *Coherence in Thought and Action*; Paul Thagard, *Hot Thought: Mechanisms and Applications of Emotional Cognition* (Cambridge, MA: MIT Press, 2006).
10. Aristotle, *The Complete Works of Aristotle*, ed. J. Barnes, 2 vols. (Princeton, NJ: Princeton University Press, 1984).
11. Michael R. DePaul, *Balance and Refinement: Beyond Coherence Methods of Moral Inquiry* (Milton Park, UK: Routledge, 1993); Patricia Marino, *Moral Reasoning in a Pluralistic World* (Montreal, QC: McGill-Queen's University Press, 2015); Mordechai Nisan, "The Moral Balance Model: Theory and Research Extending Our Understanding of Moral Choice and Deviation," in *Handbook of Moral Behavior and Development: Research*, ed. William Kurtines and J. L. Gewirtz, 213–49 (Mahwah, NJ: Erlbaum, 1991).
12. Tom L. Beauchamp and James F. Childress, *Principles of Biomedical Ethics*, 7th ed. (New York: Oxford University Press, 2013).
13. John Rawls, *A Theory of Justice* (Cambridge, MA: Harvard University Press, 1971), 21. See also John Rawls, *Political Liberalism* (New York: Columbia University Press, 1993); Norman Daniels, "Reflective Equilibrium," *Stanford Encyclopedia of Philosophy*, 2020, https://plato.stanford.edu/entries/reflective-equilibrium/.
14. Paul Thagard, *Computational Philosophy of Science* (Cambridge, MA: MIT Press, 1988), chap. 7; Norbert Paulo, "The Unreliable Intuitions Objection Against Reflective Equilibrium," *Journal of Ethics* 24, no. 3 (2020): 333–53.
15. Thagard, *The Brain and the Meaning of Life*; Thagard, *Natural Philosophy*; Paul Thagard, *Bots and Beasts: What Makes Machines, Animals, and People Smart?* (Cambridge, MA: MIT Press, 2021).
16. Richard M. Ryan and Edward L. Deci, *Self-Determination Theory: Basic Psychological Needs in Motivation, Development, and Wellness* (New York: Guilford, 2017).
17. Thagard, *Natural Philosophy*, 13–14.

18. Albert Camus, *The Myth of Sisyphus*, trans. J. O'Brien (London: Penguin, 2000); Clara E. Hill, *Meaning in Life: A Therapist's Guide* (Washington, DC: American Psychological Association, 2018); Iddo Landau, *Finding Meaning in an Imperfect World* (New York: Oxford University Press, 2017); Thaddeus Metz, *Meaning in Life: An Analytic Study* (Oxford: Oxford University Press, 2013); Paul Thagard, "The Relevance of Neuroscience to Meaning in Life," in *Oxford Handbook of Meaning in Life*, ed. Iddo Landau (Oxford: Oxford University Press, forthcoming).
19. John F. Helliwell, Richard Layard, Jeffrey Sachs, and Jan-Emmanuel De Neve, eds., *World Happiness Report 2021* (New York: Sustainable Development Solutions Network, 2021).
20. Thagard, *Brain-Mind.*
21. L. Wittgenstein, *Philosophical Investigations*, trans. G. E. M. Anscombe, 2nd ed. (Oxford: Blackwell, 1968), #124.

BIBLIOGRAPHY

Abboud, François M. "In Search of Autonomic Balance: The Good, the Bad, and the Ugly." *American Journal of Physiology: Regulatory, Integrative, and Comparative Physiology* 298, no. 6 (2010): R1449–67.

Ainslie, George. "Willpower with and Without Effort." *Behavioral and Brain Sciences* 44 (2021).

Alpert, Patricia T., Sally K. Miller, Harvey Wallmann, Richard Havey, Chad Cross, Theresa Chevalia, Carrie B. Gillis, and Keshavan Kodandapari. "The Effect of Modified Jazz Dance on Balance, Cognition, and Mood in Older Adults." *Journal of the American Academy of Nurse Practitioners* 21, no. 2 (2009): 108–15.

Alsalman, Ola, Jan Ost, Robby Vanspauwen, Catherine Blaivie, Dirk De Ridder, and Sven Vanneste. "The Neural Correlates of Chronic Symptoms of Vertigo Proneness in Humans." *PLoS One* 11, no. 4 (2016): e0152309.

Angelaki, Dora E., and Kathleen E. Cullen. "Vestibular System: The Many Facets of a Multimodal Sense." *Annual Review of Neuroscience* 31 (2008): 125–50.

Aristotle. *The Complete Works of Aristotle.* Ed. J. Barnes. 2 vols. Princeton, NJ: Princeton University Press, 1984.

Arnheim, Rudolf. *Art and Visual Perception: The New Version.* Berkeley: University of California Press, 1974.

Auilier, Dan. *Vertigo: The Making of a Hitchcock Classic.* Seattle: Amazon Kindle, 2013.

Ball, Philip. "Neuroscience Readies for a Showdown over Consciousness Ideas." *Quanta Magazine*, 2019. https://www.quantamagazine.org/neuroscience-readies-for-a-showdown-over-consciousness-ideas-20190306/.

Barbey, Aron K. "Network Neuroscience Theory of Human Intelligence." *Trends in Cognitive Sciences* 22, no. 1 (2018): 8–20.

Barrett, Lisa F., and Ajay B. Satpute. "Large-Scale Brain Networks in Affective and Social Neuroscience: Towards an Integrative Functional Architecture of the Brain." *Current Opinion in Neurobiology* 23, no. 3 (2013): 361–72.

Barsalou, Lawrence W. "Perceptual Symbol Systems." *Behavioral and Brain Sciences* 22 (1999): 577–660.

——. "Simulation, Situated Conceptualization, and Prediction." *Philosophical Transactions of the Royal Society B* 364 (2009): 1218–89.

Beaty, Roger E., Mathias Benedek, Scott Barry Kaufman, and Paul J. Silvia. "Default and Executive Network Coupling Supports Creative Idea Production." *Scientific Reports* 5 (2015): 10964.

Beauchamp, Tom L., and James F. Childress. *Principles of Biomedical Ethics*. 7th ed. New York: Oxford University Press, 2013.

Bechtel, William. *Mental Mechanisms: Philosophical Perspectives on Cognitive Neuroscience*. New York: Routledge, 2008.

Beinhocker, Eric D. *The Origin of Wealth: Evolution, Complexity, and the Radical Remaking of Economics*. Cambridge, MA: Harvard Business School Press, 2006.

Bisdorff, Alexandre, Gilles Bosser, René Gueguen, and Philippe Perrin. "The Epidemiology of Vertigo, Dizziness, and Unsteadiness and Its Links to Co-Morbidities." *Frontiers in Neurology* 4 (2013): 29.

Bisdorff, Alexandre, Michael Von Brevern, Thomas Lempert, and David E. Newman-Toker. "Classification of Vestibular Symptoms: Towards an International Classification of Vestibular Disorders." *Journal of Vestibular Research* 19, nos. 1–2 (2009): 1–13.

Blouw, Peter, Eugene Solodkin, Paul Thagard, and Chris Eliasmith. "Concepts as Semantic Pointers: A Framework and Computational Model." *Cognitive Science* 40 (2016): 1128–62.

Boisgontier, Mattieu P., Boris Cheval, Sima Chalavi, Peter van Ruitenbeek, Inge Leunissen, Oron Levin, Alice Nieuwboer, and Stephan P. Swinnen. "Individual Differences in Brainstem and Basal Ganglia Structure Predict Postural Control and Balance Loss in Young and Older Adults." *Neurobiology of Aging* 50 (2017): 47–59.

Born, Richard T., and David C. Bradley. "Structure and Function of Visual Area MT." *Annual Review of Neuroscience* 28 (2005): 157–89.

Borst, Alexander, and Thomas Euler. "Seeing Things in Motion: Models, Circuits, and Mechanisms." *Neuron* 71, no. 6 (2011): 974–94.

Bowdle, Brian F, and Dedre Gentner. "The Career of Metaphor." *Psychological Review* 112, no. 1 (2005): 193–216.

Brandt, Thomas. *Vertigo: Its Multisensory Syndromes.* London: Springer, 2003.

Bremmer, Frank, Michael Kubischik, Martin Pekel, Klaus-Peter Hoffmann, and Markus Lappe. "Visual Selectivity for Heading in Monkey Area MST." *Experimental Brain Research* 200, no. 1 (2010): 51–60.

Brissenden, James A., Emily J. Levin, David E. Osher, Mark A. Halko, and David C. Somers. "Functional Evidence for a Cerebellar Node of the Dorsal Attention Network." *Journal of Neuroscience* 36, no. 22 (2016): 6083–96.

Brown, S. Max, and Tanveer Naseer. *Leadership Vertigo: Why Even the Best Leaders Go Off Course and How They Can Get Back on Track.* Sanger, CA: Familius, 2014.

Buckner, Randy L., and Fenna M. Krienen. "The Evolution of Distributed Association Networks in the Human Brain." *Trends in Cognitive Sciences* 17, no. 12 (2013): 648–65.

Burns, Joseph C., and Jennifer S. Stone. "Development and Regeneration of Vestibular Hair Cells in Mammals." *Seminars in Cell and Developmental Biology* 65 (2017): 96–105.

Butterfield, Herbert. "Balance of Power." In *Dictionary of the History of Ideas*, ed. P. P. Wiener, 179–88. New York: Scribner's, 1973.

Bynum, William F., Anne Hardy, Stephen Jacyna, Christopher Lawrence, and E. M. Tansey. *The Western Medical Tradition: 1800–2000.* Cambridge: Cambridge University Press, 2006.

Camus, Albert. *The Myth of Sisyphus.* Trans. J. O'Brien. London: Penguin, 2000.

Cardillo, Eileen R., Christine E. Watson, Gwenda L. Schmidt, Alexander Kranjec, and Anjan Chatterjee. "From Novel to Familiar: Tuning the Brain for Metaphors." *Neuroimage* 59, no. 4 (2012): 3212–21.

Carter, Olivia, Jakob Hohwy, Jeroen Van Boxtel, Victor Lamme, Ned Block, Christof Koch, and Naotsugu Tsuchiya. "Conscious Machines: Defining Questions." *Science* 359, no. 6374 (2018): 400.

Chalmers, David J. *The Conscious Mind.* Oxford: Oxford University Press, 1996.

Chemero, Anthony. *Radical Embodied Cognitive Science.* Cambridge, MA: MIT Press, 2009.

Chiu, Chiu-Wen, Tsung-Chieh Lee, Po-Chi Hsu, Chia-Yun Chen, Shun-Chang Chang, John Y. Chiang, and Lun-Chien Lo. "Efficacy and Safety of Acupuncture for Dizziness and Vertigo in Emergency Department: A Pilot Cohort Study." *BMC Complementary and Alternative Medicine* 15 (2015): 173.

Chowdhury, Syed A., Katsumasa Takahashi, Gregory C. DeAngelis, and Dora E. Angelaki. "Does the Middle Temporal Area Carry Vestibular Signals Related to Self-Motion?" *Journal of Neuroscience* 29, no. 38 (2009): 12020–30.

Christensen, Clayton M., Michael E. Raynor, and Rory McDonald. "What Is Disruptive Innovation?" *Harvard Business Review* 93, no. 12 (2015): 44–53.

Clarke, Christopher E. "A Question of Balance: The Autism-Vaccine Controversy in the British and American Elite Press." *Science Communication* 30, no. 1 (2008): 77–107.

Collins, Randall. *Interaction Ritual Chains.* Princeton, NJ: Princeton University Press, 2004.

Conrad, Lawrence I., Michael Neve, Vivian Nutton, Roy Porter, and Andrew Wear. *The Western Medical Tradition: 800 BC to AD 1800.* Cambridge: Cambridge University Press, 1995.

Cooper, Gregory. "Must There Be a Balance of Nature?" *Biology and Philosophy* 16, no. 4 (2001): 481–506.

Craver, Carl F., and Lindley Darden. *In Search of Mechanisms: Discoveries Across the Life Sciences.* Chicago: University of Chicago Press, 2013.

Crawford, Eric, Matthew Gingerich, and Chris Eliasmith. "Biologically Plausible, Human-Scale Knowledge Representation." *Cognitive Science* 40 (2016): 782–821.

Crick, Francis. *The Astonishing Hypothesis: The Scientific Search for the Soul.* London: Simon and Schuster, 1994.

Cuddington, Kim. "The 'Balance of Nature' Metaphor and Equilibrium in Population Ecology." *Biology and Philosophy* 16, no. 4 (2001): 463–79.

Cullen, Kathleen E. "The Vestibular System: Multimodal Integration and Encoding of Self-Motion for Motor Control." *Trends in Neurosciences* 35, no. 3 (2012): 185–96.

Damasio, Anthony, and Gil B. Carvalho. "The Nature of Feelings: Evolutionary and Neurobiological Origins." *Nature Reviews Neuroscience* 14, no. 2 (2013): 143–52.

Dammann, Olaf. *Etiological Explanations.* Boca Raton, FL: CRC, 2020.

Dammann, Olaf, Ted Poston, and Paul Thagard. "How Do Medical Researchers Make Causal Inferences?" In *What Is Scientific Knowledge? An Introduction to Contemporary Epistemology of Science*, ed. K. McCain and K. Kampourakis, 33–51. New York: Routledge, 2019.

Daniels, Norman. "Reflective Equilibrium." *Stanford Encyclopedia of Philosophy*, 2020. https://plato.stanford.edu/entries/reflective-equilibrium/.

Darden, Lindley, Lipika R. Pal, Kunal Kundu, and John Moult. "The Product Guides the Process: Discovering Disease Mechanisms." In *Building Theories: Heuristics and Hypotheses in the Sciences*, ed. David Danks and Emiliano Ippoliti, 101–17. Cham, Switzerland: Springer, 2018.

Dehaene, Stanislas. *Consciousness and the Brain: Deciphering How the Brain Codes Our Thoughts*. New York: Viking, 2014.

Dehaene, Stanislas, Hakwan Lau, and Sid Kouider. "What Is Consciousness, and Could Machines Have It?" *Science* 358, no. 6362 (2017): 486–92.

DePaul, Michael R. *Balance and Refinement: Beyond Coherence Methods of Moral Inquiry*. London: Routledge, 1993.

Diamond, Jared. *Collapse: How Societies Choose to Fail or Succeed*. New York: Penguin, 2011.

Diekhof, Esther K., and Oliver Gruber. "When Desire Collides with Reason: Functional Interactions Between Anteroventral Prefrontal Cortex and Nucleus Accumbens Underlie the Human Ability to Resist Impulsive Desires." *Journal of Neuroscience* 30, no. 4 (2010): 1488–93.

Dieterich, Marianne, and Thomas Brandt. "Functional Brain Imaging of Peripheral and Central Vestibular Disorders." *Brain* 131, no. 10 (2008): 2538–52.

Dreyfus, Hubert L. "Why Heideggerian AI Failed and How Fixing It Would Require Making It More Heideggerian." *Philosophical Psychology* 20 (2007): 247–68.

Edwards, A. J. C. *Nuclear Weapons: The Balance of Terror, the Quest for Peace*. Albany: State University of New York Press, 1986.

Egerton, Frank N. "Changing Concepts of the Balance of Nature." *Quarterly Review of Biology* 48, no. 2 (1973): 322–50.

Eliasmith, Chris. *How to Build a Brain: A Neural Architecture for Biological Cognition*. Oxford: Oxford University Press, 2013.

Eliasmith, Chris, and Charles H. Anderson. *Neural Engineering: Computation, Representation and Dynamics in Neurobiological Systems*. Cambridge, MA: MIT Press, 2003.

Ernst, Edzard. "Acupuncture: What Does the Most Reliable Evidence Tell Us?" *Journal of Pain Symptom Management* 37, no. 4 (2009): 709–14.

Fauconnier, Gilles, and Mark Turner. *The Way We Think.* New York: Basic Books, 2002.

Festinger, Leon. *A Theory of Cognitive Dissonance.* Stanford, CA: Stanford University Press, 1957.

Foin, Theodore C., and William G. Davis. "Equilibrium and Nonequilibrium Models in Ecological Anthropology: An Evaluation of 'Stability' in Maring Ecosystems in New Guinea." *American Anthropologist* 89, no. 1 (1987): 9–31.

Forest, Amanda L., David R. Kille, Joanne V. Wood, and Lindsay R. Stehouwer. "Turbulent Times, Rocky Relationships: Relational Consequences of Experiencing Physical Instability." *Psychological Sciences* 26, no. 8 (2015): 1261–71.

Franklin, James. *An Aristotelian Realist Philosophy of Mathematics.* Houndmills, UK: Palgrave Macmillan, 2014.

Frejo, Lidia, Ina Giegling, Robert Teggi, Jose A. Lopez-Escamez, and Dan Rujescu. "Genetics of Vestibular Disorders: Pathophysiological Insights." Supplement, *Journal of Neurology, Neurosurgery and Psychiatry* 263, no. S1 (2016): S45–53.

Furman, Joseph M., and Thomas Lempert. *Neuro-Otology.* Amsterdam: Elsevier, 2016.

Gallagher, Shaun. *How the Body Shapes the Mind.* Oxford: Oxford University Press, 2006.

Gazquez, Irene, and Jose A. Lopez-Escamez. "Genetics of Recurrent Vertigo and Vestibular Disorders." *Current Genomics* 12, no. 6 (2011): 443–50.

Gentner, Dedre, and Brian Bowdle. "Metaphor as Structure-Mapping." In *The Cambridge Handbook of Metaphor and Thought*, ed. R. W. Gibbs, 109–28. Cambridge: Cambridge University Press, 2008.

Gerken, Mikkel. "How to Balance Balanced Reporting and Reliable Reporting." *Philosophical Studies* 177, no. 10 (2020): 3117–42.

Gerritsen, Roderik J. S., and Guido P. H. Band. "Breath of Life: The Respiratory Vagal Stimulation Model of Contemplative Activity." *Frontiers in Human Neuroscience* 12 (2018): 397.

Gibbs, Ray W. *Embodiment and Cognitive Science.* Cambridge: Cambridge University Press, 2005.

——. *Metaphor Wars.* Cambridge: Cambridge University Press, 2017.

Gladwell, Malcolm. *The Tipping Point: How Little Things Can Make a Big Difference.* New York: Little, Brown, 2006.

Glennan, Stuart. *The New Mechanical Philosophy.* Oxford: Oxford University Press, 2017.

Glimcher, Paul W. "Understanding Dopamine and Reinforcement Learning: The Dopamine Reward Prediction Error Hypothesis." Supplement, *Proceedings of the National Academy of Sciences* 108, no. S3 (2011): S15647–54.

Goff, Philip, William Seager, and Sean Allen-Hermanson. "Panpsychism." *Stanford Encyclopedia of Philosophy*, 2017. https://plato.stanford.edu/entries/panpsychism/.

Goldberg, Jay M., Victor J. Wilson, Dora E. Angelaki, Kathleen E. Cullen, and Kikuro Fukushima. *The Vestibular System: A Sixth Sense.* Oxford: Oxford University Press, 2012.

Goldman, Alvin. *Epistemology and Cognition.* Cambridge, MA: Harvard University Press, 1986.

Gould, Stephen Jay, and Niles Eldredge. "Punctuated Equilibrium Comes of Age." *Nature* 366, no. 6452 (1993): 223–27.

Green, Andrea M., and Dora E. Angelaki. "Internal Models and Neural Computation in the Vestibular System." *Experimental Brain Research* 200, nos. 3–4 (2010): 197–222.

Gregory, Richard Langton. "Perceptions as Hypotheses." *Philosophical Transactions of the Royal Society of London B, Biological Sciences* 290, no. 1038 (1980): 181–97.

Grodzins, Morton. "Metropolitan Segregation." *Scientific American* 197, no. 4 (1957): 33–41.

Guerra, Joaquín, and Ramón Cacabelos. "Pharmacoepigenetics of Vertigo and Related Vestibular Syndromes." In *Pharmacoepigenetics*, ed. Ramón Cacabelos, 755–79. Cambridge, MA: Academic, 2019.

Haack, Susan. *Evidence and Inquiry: Towards Reconstruction in Epistemology.* Oxford: Blackwell, 1993.

Hain, Timothy C., Todd M. Squires, and Howard A. Stone. "Clinical Implications of a Mathematical Model of Benign Paroxysmal Positional Vertigo." *Annals of the New York Academy of Sciences* 1039 (2005): 384–94.

Harmon-Jones, Eddie, ed. *Cognitive Dissonance: Reexamining a Pivotal Theory in Psychology.* Washington, DC: American Psychological Association, 2019.

Harris, Randy Allen. *The Linguistic Wars*. 2nd ed. Oxford: Oxford University Press, 2021.

Hausman, Daniel M. *The Inexact and Separate Science of Economics*. Cambridge: Cambridge University Press, 1992.

Healy, David. "Serotonin and Depression." *BMJ* 350 (2015): h1771.

Heider, Fritz. "Attitudes and Cognitive Organization." *Journal of Psychology* 21, no. 1 (1946): 107–12.

——. *The Psychology of Interpersonal Relations*. New York: Wiley, 1958.

Helliwell, John F., Richard Layard, Jeffrey Sachs, and Jan-Emmanuel De Neve, eds. *World Happiness Report 2021*. New York: Sustainable Development Solutions Network, 2021.

Hill, Clara E. *Meaning in Life: A Therapist's Guide*. Washington DC: American Psychological Association, 2018.

Hills, David. "Metaphor." *Stanford Encyclopedia of Philosophy* 2016. https://plato.stanford.edu/entries/metaphor/.

Hippocrates. *Delphi Complete Works of Hippocrates*. East Sussex, UK: Delphi Classics, 2015.

Holyoak, Keith J. "Analogy and Relational Reasoning." In *The Oxford Handbook of Thinking and Reasoning*, ed. Keith. J. Holyoak and Robert G. Morrison, 234–59. Oxford: Oxford University Press, 2012.

——. *The Spider's Thread: Metaphor in Mind, Brain, and Poetry*. Cambridge, MA: MIT Press, 2019.

Holyoak, Keith J., and Paul Thagard. *Mental Leaps: Analogy in Creative Thought*. Cambridge, MA: MIT Press, 1995.

Hrysomallis, Con. "Balance Ability and Athletic Performance." *Sports Medicine* 41, no. 3 (2011): 221–32.

Huang, Zirui, Jun Zhang, Jinsong Wu, George A. Mashour, and Anthony G. Hudetz. "Temporal Circuit of Macroscale Dynamic Brain Activity Supports Human Consciousness." *Science Advances* 6, no. 11 (2020): eaaz0087.

Hume, David. *Enquiries Concerning Human Understanding and Concerning the Principles of Morals*. Ed. L. A. Selby-Bigge. 2nd ed. Oxford: Oxford University Press, 1972.

Huston, Patricia, and Bruce McFarlane. "Health Benefits of Tai Chi: What Is the Evidence?" *Canadian Family Physician* 62, no. 11 (2016): 881–90.

Hutcheson, Francis. *An Inquiry into the Original of Our Ideas of Beauty and Virtue: In Two Treatises*. London: R. Ware, 1753.

Hutt, Kimberley, and Emma Redding. "The Effect of an Eyes-Closed Dance-Specific Training Program on Dynamic Balance in Elite Pre-Professional Ballet Dancers: A Randomized Controlled Pilot Study." *Journal of Dance Medicine and Science* 18, no. 1 (2014): 3–11.

Ilg, Uwe J. "The Role of Areas MT and MST in Coding of Visual Motion Underlying the Execution of Smooth Pursuit." *Vision Research* 48, no. 20 (2008): 2062–69.

Iwasaki, Shinichi, and Tatsuya Yamasoba. "Dizziness and Imbalance in the Elderly: Age-Related Decline in the Vestibular System." *Aging and Disease* 6, no. 1 (2015): 38–47.

Jelinski, Dennis E. "There Is No Mother Nature—There Is No Balance of Nature: Culture, Ecology and Conservation." *Human Ecology* 33, no. 2 (2005): 271–88.

Jeter, Pamela E., Amélie-Françoise Nkodo, Steffany H. Moonaz, and Gislin Dagnelie. "A Systematic Review of Yoga for Balance in a Healthy Population." *Journal of Alternative and Complementary Medicine* 20, no. 4 (2014): 221–32.

Johnson, Mark. *The Body in the Mind.* Chicago: University of Chicago Press, 1987.

Kahneman, Daniel, and Amos Tversky, eds. *Choices, Values, and Frames.* Cambridge: Cambridge University Press, 2000.

Kajić, Ivana, Tobias C. Schröder, Terrence C. Stewart, and Paul Thagard. "The Semantic Pointer Theory of Emotions." *Cognitive Systems Research* 58 (2019): 35–53.

Kandel, Eric R., James H. Schwartz, Thomas M. Jessell, Steven Siegelbaum, and A. J. Hudspeth. *Principles of Neural Science.* 5th ed. New York: McGraw-Hill, 2012.

Karim, Helmet T., Patrick J. Sparto, Howard J. Aizenstein, Joseph M. Furman, Theodore J. Huppert, Kirk I. Erickson, and Patrick J. Loughlin. "Functional MR Imaging of a Simulated Balance Task." *Brain Research* 1555 (2014): 20–27.

Kelz, Max B., and George A. Mashour. "The Biology of General Anesthesia from Paramecium to Primate." *Current Biology* 29, no. 22 (2019): R1199–210.

Ketabi, Shara Rose. *Ayurveda.* Indianapolis, IN: Dorling Kindersley, 2017.

Kille, David R., Amanda L. Forest, and Joanne V. Wood. "Tall, Dark, and Stable: Embodiment Motivates Mate Selection Preferences." *Psychological Science* 24, no. 1 (2013): 112–14.

Koch, Christof. *The Feeling of Life Itself: Why Consciousness Is Widespread but Can't Be Computed.* Cambridge, MA: MIT Press, 2019.

Kolbert, Elizabeth. *The Sixth Extinction: An Unnatural History.* London: A&C Black, 2014.

Kricher, John. *The Balance of Nature: Ecology's Enduring Myth.* Princeton, NJ: Princeton University Press, 2009.

Kshirsagar, Manisha, and Ana C. Magno. *Ayurveda: A Quick Reference Handbook.* Twin Lakes, WI: Lotus, 2011.

Kunda, Ziva. "The Case for Motivated Reasoning." *Psychological Bulletin* 108 (1990): 480–98.

Lacasse, Jeffrey R., and Jonathan Leo. "Antidepressants and the Chemical Imbalance Theory of Depression: A Reflection and Update on the Discourse." *The Behavior Therapist* 38, no. 7 (2015): 206–13.

Lakoff, George, and Mark Johnson. *Metaphors We Live By.* Chicago: University of Chicago Press, 1980.

——. *Philosophy in the Flesh: The Embodied Mind and Its Challenge to Western Thought.* New York: Basic Books, 1999.

Lakoff, George, and Rafael E. Núñez. *Where Mathematics Comes From: How the Embodied Mind Brings Mathematics into Being.* New York: Basic Books, 2000.

Landau, Iddp. *Finding Meaning in an Imperfect World.* New York: Oxford University Press, 2017.

Landau, Mark J. *Conceptual Metaphor in Social Psychology: The Poetics of Everyday Life.* New York: Routledge, 2017.

Lenton, Timothy M., Johan Rockström, Owen Gaffney, Stefan Rahmstorf, Katherine Richardson, Will Steffen, and Hans Joachim Schellnhuber. "Climate Tipping Points—Too Risky to Bet Against." *Nature* 575 (2019): 592–95.

Levitin, Daniel J. *Successful Aging: A Neuroscientist Explores the Power and Potential of Our Lives.* New York: Dutton, 2020.

Li, Yonghui, Yong Liu, Jun Li, Wen Qin, Kuncheng Li, Chunsui Yu, and Tinazi Jiang. "Brain Anatomical Network and Intelligence." *PLoS Computational Biology* 5, no. 5 (2009): e1000395.

Lindauer, Maurice W. "The Evolution of the Concept of Chemical Equilibrium from 1775 to 1923." *Journal of Chemical Education* 39, no. 8 (1962): 384.

Liu, Shenbin, Zhi-Fu Wang, Yuang-Shai Su, Russell S. Ray, Xiang-Hong Jing, Yan-Qing Wang, and Quifu Ma. "Somatotopic Organization and Intensity Dependence in Driving Distinct Npy-Expressing Sympathetic Pathways by Electroacupuncture." *Neuron* 108, no. 3 (2020): 436–50.e7.

Lopez, Christophe. "A Neuroscientific Account of How Vestibular Disorders Impair Bodily Self-Consciousness." *Frontiers in Integrative Neuroscience* 7 (2013): 91.

Lopez, Christophe, and Olaf Blanke. "The Thalamocortical Vestibular System in Animals and Humans." *Brain Research Reviews* 67, nos. 1–2 (2011): 119–46.

Luthar, Suniya S., Emily L. Lyman, and Elizabeth J. Crossman. "Resilience and Positive Psychology." In *Handbook of Developmental Psychopathology*, ed. M. Lewis and K. D. Rudolph, 125–40. New York: Springer, 2014.

Magnani, Lorenzo. *Abductive Cognition: The Epistemological and Eco-Cognitive Dimensions of Hypothetical Reasoning.* Berlin: Springer, 2009.

Manheimer, Eric, Susan Wieland, Elizabeth Kimbrough, Ker Cheng, and Brian M. Berman. "Evidence from the Cochrane Collaboration for Traditional Chinese Medicine Therapies." *Journal of Alternative and Complementary Medicine* 15, no. 9 (2009): 1001–14.

Margolis, J. Eli. "Understanding Political Stability and Instability." *Civil Wars* 12, no. 3 (2010): 326–45.

Marino, Patricia. *Moral Reasoning in a Pluralistic World.* Montreal, QC: McGill-Queen's University Press, 2015.

Marr, David, and Tomaso Poggio. "Cooperative Computation of Stereo Disparity." *Science* 194 (1976): 283–87.

Marshall, Alfred. *Principles of Economics.* London: Macmillan, 1890.

——. *Principles of Economics*. 8th ed. London: Macmillan, 1920.

Martin, Roger L. *When More Is Not Better: Overcoming America's Obsession with Economic Efficiency.* Boston: Harvard Business Review Press, 2020.

Mashour, George A., and Anthony G. Hudetz. "Neural Correlates of Unconsciousness in Large-Scale Brain Networks." *Trends in Neurosciences* 41, no. 3 (2018): 150–60.

Mason, Robert A., and Marcel Adam Just. "Neural Representations of Procedural Knowledge." *Psychological science* 31, no. 6 (2020): 729–40.

McClelland, James L., Daniel Mirman, Donald J. Bolger, and Pranav Khaitan. "Interactive Activation and Mutual Constraint Satisfaction in Perception and Cognition." *Cognitive Science* 38, no. 6 (2014): 1139–89.

McClure, Samuel M., David I. Laibson, George Loewenstein, and Jonathan D. Cohen. "Separate Neural Systems Value Immediate and Delayed Monetary Rewards." *Science* 306, no. 5695 (2004): 503–7.

Melloni, Lucia, Liad Mudrik, Michael Pitts, and Christof Koch. "Making the Hard Problem of Consciousness Easier." *Science* 372, no. 6545 (2021): 911–12.

Merker, Bjorn. "Consciousness Without a Cerebral Cortex: A Challenge for Neuroscience and Medicine." *Behavioral and Brain Sciences* 30, no. 1 (2007): 63–81; discussion 81–134.

Metz, Thaddeus. *Meaning in Life: An Analytic Study.* Oxford: Oxford University Press, 2013.

Milkoreit, Manjana, Jennifer Hodbod, Jacopo Baggio, Karina Benessaiah, Rafael Calderón-Contreras, Jonathan F. Donges, Jean-Denis Mathias, et al. "Defining Tipping Points for Social-Ecological Systems Scholarship—An Interdisciplinary Literature Review." *Environmental Research Letters* 13, no. 3 (2018).

Miller, George A. "The Magical Number Seven, Plus or Minus Two: Some Limits on Our Capacity for Processing Information." *Psychological Review* 63 (1956): 81–97.

Misiaszek, John E. "Neural Control of Walking Balance: If Falling Then React Else Continue." *Exercise and Sport Sciences Reviews* 34, no. 3 (2006): 128–34.

Mistry, Rohinton. *A Fine Balance.* Toronto, ON: McClelland and Stewart, 1997.

Murphy, Dominic. "Concepts of Disease and Health." *Stanford Encyclopedia of Philosophy*, 2020. https://plato.stanford.edu/entries/health-disease/.

Nichols, M. James, and William T. Newsome. "Middle Temporal Visual Area Microstimulation Influences Veridical Judgments of Motion Direction." *Journal of Neuroscience* 22, no. 21 (2002): 9530–40.

Nisan, Mordechai. "The Moral Balance Model: Theory and Research Extending Our Understanding of Moral Choice and Deviation." In *Handbook of Moral Behavior and Development: Research*, ed. William Kurtines and J. L. Gewirtz, 213–49. Mahwah, NJ: Erlbaum, 1991.

Ordeshook, Peter C., and Kenneth A. Shepsle. *Political Equilibrium: A Delicate Balance.* Boston: Kluwer-Nijhoff, 1982.

Ortony, Andrew, ed. *Metaphor and Thought.* Cambridge: Cambridge University Press, 1993.

Paley, Carol A., and Mark I. Johnson. "Acupuncture for the Relief of Chronic Pain: A Synthesis of Systematic Reviews." *Medicina* 56 (2019): 1.

Parkkinen, Veli-Pekka, Christian Wallmann, Michael Wilde, Brendan Clarke, Phyllis Illari, Michael P. Kelly, Charles Norell, et al. *Evaluating Evidence of Mechanisms in Medicine: Principles and Procedures.* Berlin: Springer Nature, 2018.

Parnes, Lorne S., Sumit K. Agrawal, and Jason Atlas. "Diagnosis and Management of Benign Paroxysmal Positional Vertigo (Bppv)." *Canadian Medical Association Journal* 169, no. 7 (2003): 681–93.

Parsons, Talcott. *The Structure of Social Action.* New York: Free Press, 1949.

Patwardhan, Bhushan, Dnyaneshwar Warude, Palpu Pushpangadan, and Narendra Bhatt. "Ayurveda and Traditional Chinese Medicine: A Comparative Overview." *Evidence-Based Complementary and Alternative Medicine* 2, no. 4 (2005): 465–73.

Paulo, Norbert. "The Unreliable Intuitions Objection Against Reflective Equilibrium." *Journal of Ethics* 24, no. 3 (2020): 333–53.

Petruso, Karl M. "Early Weights and Weighing in Egypt and the Indus Valley." *M Bulletin (Museum of Fine Arts, Boston)* 79 (1981): 44–51.

Pfeiffer, Christian, Andrew Serino, and Olaf Blanke. "The Vestibular System: A Spatial Reference for Bodily Self-Consciousness." *Frontiers in Integrative Neuroscience* 8 (2014): 31.

Pimm, Stuart L. *The Balance of Nature: Ecological Issues in the Conservation of Species and Communities.* Chicago: University of Chicago Press, 1991.

Pippin, Robert B. *The Philosophical Hitchcock: "Vertigo" and the Anxieties of Unknowingness.* Chicago: University of Chicago Press, 2017.

Politi, Daniel. "Richard Thaler Wins Economics Nobel for Recognizing People Are Irrational." *Slate*, 2017. http://www.slate.com/blogs/the_slatest/2017/10/09/richard_thaler_wins_economics_nobel_for_recognizing_that_people_are_irrational.html.

Porkert, Manfred, and Christian Ullmann. *Chinese Medicine.* Trans. M. Howson. New York: Morrow, 1988.

Post, Robert E., and Lori M. Dickerson. "Dizziness: A Diagnostic Approach." *American Family Physician* 82, no. 4 (2010): 361–68.

Puig, Berta, Sandra Brenna, and Tim Magnus. "Molecular Communication of a Dying Neuron in Stroke." *International Journal of Molecular Sciences* 19, no. 9 (2018).

Raichle, Marcus E. "The Brain's Default Mode Network." *Annual Review of Neuroscience* 38 (2015): 433–47.

Rajagopalan, Archana, K. V. Jinu, Kumar Sai Sailesh, Soumya Mishra, Udaya Kumar Reddy, and Joseph Kurien Mukkadan. "Understanding the Links Between Vestibular and Limbic Systems Regulating Emotions." *Journal of Natural Science, Biology and Medicine* 8, no. 1 (2017): 11–15.

Ranney, Michael A., and Dav Clark. "Climate Change Conceptual Change: Scientific Information Can Transform Attitudes." *Topics in Cognitive Science* 8, no. 1 (2016): 49–75.

Rauch, Steven D., Luis Velazquez-Villaseñor, Paul S. Dimitri, and Saumil N. Merchant. "Decreasing Hair Cell Counts in Aging Humans." *Annals of the New York Academy of Sciences* 942, no. 1 (2001): 220–27.

Rawls, John. *Political Liberalism.* New York: Columbia University Press, 1993.

——. *A Theory of Justice.* Cambridge, MA: Harvard University Press, 1971.

Read, Stephen J., Eric J. Vanman, and Lynn C. Miller. "Connectionism, Parallel Constraint Satisfaction, and Gestalt Principles: (Re)Introducing Cognitive Dynamics to Social Psychology." *Personality and Social Psychology Review* 1 (1997): 26–53.

Renn, Jürgen, and Peter Damerow. *The Equilibrium Controversy: Guidobaldo Del Monte's Critical Notes on the Mechanics of Jordanus and Benedetti and Their Historical and Conceptual Backgrounds.* Berlin: Edition Open Access, 2017.

Roberts, W. M., J. Howard, and A. J. Hudspeth. "Hair Cells: Transduction, Tuning, and Transmission in the Inner Ear." *Annual Review of Cell Biology* 4, no. 1 (1988): 63–92.

Robinson, Howard. "Dualism." *Stanford Encyclopedia of Philosophy*, 2016. https://plato.stanford.edu/entries/dualism/.

Rogge, Ann-Kathrin, Brigitte Roder, Astrid Zech, Volker Nagel, Karsten Hollander, Klaus-Michael Braumann, and Kirsten Hotting. "Balance Training Improves Memory and Spatial Cognition in Healthy Adults." *Scientific Reports* 7, no. 1 (2017): 5661.

Ryan, Richard M., and Edward L. Deci. *Self-Determination Theory: Basic Psychological Needs in Motivation, Development, and Wellness.* New York: Guilford, 2017.

Sanchez, Katherine, and Fiona J. Rowe. "Role of Neural Integrators in Oculomotor Systems: A Systematic Narrative Literature Review." *Acta Ophthalmologica* 96, no. 2 (2018): e111–18.

Sartre, Jean-Paul. *Nausea.* Trans. L. Alexander. New York: New Directions, 1964.

Scheffer, Marten. *Critical Transitions in Nature and Society.* Princeton, NJ: Princeton University Press, 2009.

Schumpeter, Joseph A. *Capitalism, Socialism and Democracy.* New York: Harper, 1942.

Shultz, Thomas R., and Mark R. Lepper. "Cognitive Dissonance Reduction as Constraint Satisfaction." *Psychological Review* 103 (1996): 219–40.

Smith, Edward E., and Stephen M. Kosslyn. *Cognitive Psychology: Mind and Brain.* Upper Saddle River, NJ: Pearson Prentice Hall, 2007.

Sontag, Susan. *Illness as Metaphor.* New York: Vintage, 1979.

Sporns, Olaf. "The Human Connectome: Origins and Challenges." *Neuroimage* 80 (2013): 53–61.

Steen, Gerard J., Aletta G. Dorst, J. Berenike Herrmann, Anna A. Kaal, and Tina Krennmayr. "Metaphor in Usage." *Cognitive Linguistics* 21, no. 4 (2010): 765–96.

Stibbe, A. "The Role of Image Systems in Complementary Medicine." *Complementary Therapies in Medicine* 6, no. 4 (1998): 190–94.

Stiglitz, Joseph E. *Freefall: America, Free Markets, and the Sinking of the World Economy.* New York: Norton, 2010.

Stux, Gabriel, Brian Berman, and Bruce Pomeranz. *Basics of Acupuncture.* Berlin: Springer, 2012.

Sutter, Daniel. "The Democratic Efficiency Debate and Definitions of Political Equilibrium." *Review of Austrian Economics* 15, nos. 2–3 (2002): 199–209.

Suzuki, David. *The Sacred Balance: Rediscovering Our Place in Nature*, updated and expanded. Vancouver, BC: Greystone, 2007.

Thagard, Paul. *Bots and Beasts: What Makes Machines, Animals, and People Smart?* Cambridge, MA: MIT Press, 2021.

——. *The Brain and the Meaning of Life.* Princeton, NJ: Princeton University Press, 2010.

——. *Brain-Mind: From Neurons to Consciousness and Creativity.* New York: Oxford University Press, 2019.

——. "The Cognitive Science of Covid-19: Acceptance, Denial, and Belief Change." *Methods*, 195 (2021): 92-102.

——. *The Cognitive Science of Science: Explanation, Discovery, and Conceptual Change.* Cambridge, MA: MIT Press, 2012.

——. *Coherence in Thought and Action.* Cambridge, MA: MIT Press, 2000.

——. *Computational Philosophy of Science.* Cambridge, MA: MIT Press, 1988.

——. *Conceptual Revolutions.* Princeton, NJ: Princeton University Press, 1992.

——. "Energy Requirements Undermine Substrate Independence and Mind-Body Functionalism." *Philosophy of Science* (forthcoming).

——. "Evolution, Creation, and the Philosophy of Science." In *Evolution, Epistemology, and Science Education*, ed. R. Taylor and M. Ferrari, 20–37. Milton Park, UK: Routledge, 2010.

——. "Explanatory Coherence." *Behavioral and Brain Sciences* 12 (1989): 435–67.

——. "Explanatory Identities and Conceptual Change." *Science and Education* 23 (2014): 1531–48.

——. *Hot Thought: Mechanisms and Applications of Emotional Cognition.* Cambridge, MA: MIT Press, 2006.

——. "How Rationality Is Bounded by the Brain." In *Routledge Handbook of Bounded Rationality*, ed. Riccardo Viale, 398-406. London: Routledge, 2021.

——. *How Scientists Explain Disease.* Princeton, NJ: Princeton University Press, 1999.

——. *Mind-Society: From Brains to Social Sciences and Professions.* New York: Oxford University Press, 2019.

——. *Natural Philosophy: From Social Brains to Knowledge, Reality, Morality, and Beauty.* New York: Oxford University Press, 2019.

——. "The Relevance of Neuroscience to Meaning in Life." In *Oxford Handbook of Meaning in Life*, ed. Iddo Landau. Oxford: Oxford University Press (forthcoming).

——. "Thought Experiments Considered Harmful." *Perspectives on Science* 22 (2014): 288–305.

Thagard, Paul, and Craig Beam. "Epistemological Metaphors and the Nature of Philosophy." *Metaphilosophy* 35 (2004): 504–16.

Thagard, Paul, Laurette Larocque, and Ivana Kajić. "Emotional Change: Neural Mechanisms Based on Semantic Pointers." *Emotion* (forthcoming).

Thagard, Paul, and Terrence C. Stewart. "Two Theories of Consciousness: Semantic Pointer Competition vs. Information Integration." *Consciousness and Cognition* 30 (2014): 73–90.

Thagard, Paul, and Jing Zhu. "Acupuncture, Incommensurability, and Conceptual Change." In *Intentional Conceptual Change*, ed. G. M. Sinatra and P. R. Pintrich, 79–102. Mahwah, NJ: Erlbaum, 2003.

Thibodeau, Paul H., Rose K. Hendricks, and Lera Boroditsky. "How Linguistic Metaphor Scaffolds Reasoning." *Trends in Cognitive Sciences* 21, no. 11 (2017): 852–63.

Thompson, Timothy L., and Ronald Amedee. "Vertigo: A Review of Common Peripheral and Central Vestibular Disorders." *Ochsner Journal* 9, no. 1 (2009): 20.

Tilikete, Caroline, and Alain Vighetto. "Oscillopsia: Causes and Management." *Current Opinion in Neurology* 24, no. 1 (2011): 38–43.

Tipton, Charles M. "The History of 'Exercise Is Medicine' in Ancient Civilizations." *Advances in Physiology Education* 38, no. 2 (2014): 109–17.

Tononi, Giulio, Melanie Boly, Massimini Massimini, and Christof Koch. "Integrated Information Theory: From Consciousness to Its Physical Substrate." *Nature Reviews Neuroscience* 17, no. 7 (2016): 450–61.

Tracey, Irene, and Patrick W. Mantyh. "The Cerebral Signature for Pain Perception and Its Modulation." *Neuron* 55, no. 3 (2007): 377–91.

Tsai, Kun-Ling, Chia-To Wang, Chia-Hua Kuo, Yuan-Yang Cheng, Hsin-I Ma, Ching-Hsia Hung, Yi-Ju Tsai, and Chung-Lan Kao. "The Potential Role of Epigenetic Modulations in Bppv Maneuver Exercises." *Oncotarget* 7, no. 24 (2016): 35522.

Tversky, Amos, and Daniel Kahneman. "The Framing of Decisions and the Psychology of Choice." *Science* 211 (1981): 453–58.

Veith, Ilza. *The Yellow Emperor's Classic of Internal Medicine.* Oakland: University of California Press, 1975.

Wager, Tor D., and Lauren Y. Atlas. "The Neuroscience of Placebo Effects: Connecting Context, Learning and Health." *Nature Reviews Neuroscience* 16, no. 7 (2015): 403–18.

Wahl-Jorgensen, Karin, Mike Berry, Iñaki Garcia-Blanco, Lucy Bennett, and Jonathan Cable. "Rethinking Balance and Impartiality in Journalism? How the BBC Attempted and Failed to Change the Paradigm." *Journalism* 18, no. 7 (2017): 781–800.

Wang, Chenchen, Raveenchara Bannuru, Judith Ramel, Bruce Kupelnick, Tammy Scott, and Christopher H. Schmid. "Tai Chi on Psychological Well-Being: Systematic Review and Meta-Analysis." *BMC Complementary and Alternative Medicine* 10 (2010): 23.

Wayne, Peter M., and Mark Fuerst. *The Harvard Medical School Guide to Tai Chi: 12 Weeks to a Healthy Body, Strong Heart, and Sharp Mind.* Boston: Shambhala, 2013.

Wazen, Jack J., and Deborah Mitchell. *Dizzy: What You Need to Know About Managing and Treating Balance Disorders.* New York: Simon and Schuster, 2008.

Webb, Ben S., Timothy Ledgeway, and Francesca Rocchi. "Neural Computations Governing Spatiotemporal Pooling of Visual Motion Signals in Humans." *Journal of Neuroscience* 31, no. 13 (2011): 4917–25.

Wittgenstein, Ludwig. *Philosophical Investigations.* Trans. G. E. M. Anscombe. 2nd ed. Oxford: Blackwell, 1968.

Wu, Junjiao, and Yu Tang. "Revisiting the Immune Balance Theory: A Neurological Insight into the Epidemic of Covid-19 and Its Alike." *Frontiers in Neurology* 11 (2020): 566680.

Wulf, Gabrile. "Attentional Focus Effects in Balance Acrobats." *Research Quarterly for Exercise and Sport* 79, no. 3 (2008): 319–25.

Yates, Bill J., Michael F. Catanzaro, Daniel J. Miller, and Andrew A. McCall. "Integration of Vestibular and Emetic Gastrointestinal Signals That Produce Nausea and Vomiting: Potential Contributions to Motion Sickness." *Experimental Brain Research* 232, no. 8 (2014): 2455–69.

Yu, Angus P., Bjorn T. Tam, Christopher W. Lai, Doris S. Yu, Jean Woo, Ka-Fai Chung, Stanley S. Hui, et al. "Revealing the Neural Mechanisms Underlying the Beneficial Effects of Tai Chi: A Neuroimaging Perspective." *American Journal of Chinese Medicine* 46, no. 2 (2018): 231–59.

Ziyin, Shen, and Chen Zelin. *The Basis of Traditional Chinese Medicine.* Boston: Shambhala, 1994.

INDEX

Figures and tables are indicated by "*f*" and "*t*" after page numbers.